T0295462

Machine Vision for Industry 4.0

Machine Vision for Industry 4.0

Applications and Case Studies

Edited by

Roshani Raut
Salah-ddine Krit
Prasenjit Chatterjee

CRC Press
Taylor & Francis Group
Boca Raton London New York

CRC Press is an imprint of the
Taylor & Francis Group, an **informa** business

First edition published 2022
by CRC Press
6000 Broken Sound Parkway NW, Suite 300, Boca Raton, FL 33487-2742

and by CRC Press
2 Park Square, Milton Park, Abingdon, Oxon, OX14 4RN

Library of Congress Cataloging-in-Publication Data

Names: Raut, Roshani, 1981- editor. | Krit, Salahddine, 1976- editor. | Chatterjee,
    Prasenjit, 1982- editor.
Title: Machine vision for industry 4.0 : applications and case studies / edited by
    Roshani Raut, Salahddine Krit, and Prasenjit Chatterjee.
Description: First edition. | Boca Raton : CRC Press, [2022] | Includes bibliographical
    references and index.
Identifiers: LCCN 2021034743 | ISBN 9780367637125 (hbk) | ISBN 9780367641641
    (pbk) | ISBN 9781003122401 (ebk)
Subjects: LCSH: Computer vision—Industrial applications. | Industry 4.0.
Classification: LCC TA1634 .M33524 2022 | DDC 006.3/7—dc23
LC record available at https://lccn.loc.gov/2021034743

ISBN: 9780367637125 (hbk)
ISBN: 9780367641641 (pbk)
ISBN: 9781003122401 (ebk)

DOI: 10.1201/9781003122401

Typeset in Times
by KnowledgeWorks Global Ltd.

Dedication

The editors would like to dedicate this book to their parents, life partners, children, students, scholars, friends and colleagues.

Contents

Preface

The last two and a half centuries have produced amazing technological developments. Artificial intelligence (AI), image processing, machine vision, deep learning and analytics, software and advanced computer technology and concepts such as the Internet of Things (IoT) have propelled the digitalisation era forward. New and more advanced equipment and methodologies have arrived to push industrial capabilities to new heights, bringing about ongoing automation of traditional manufacturing and industrial practices using modern smart technology, which has emergesd as the Industry 4.0 concept that is currently gaining great research attention including writing this book. Machine Vision and IoT are combined for more automation, improved communication and self-monitoring and smart machine manufacturing, which can evaluate and resolve issues without human intervention.

The main objective of this book is to provide insights on how Machine Vision is used for Industry 4.0 for specific applications like Machine Vision in Robotics IOT, Machine Vision in the healthcare industry, Machine Vision in transportation and Machine Vision in agriculture industries. It also focuses on how smart sensors support the automation process in the different industries.

The content collected in this guidebook has been edited to provide an understanding of current research successes and problems using Machine Vision for Industry 4.0. The book targets senior and junior engineers, undergraduate and postgraduate students, research scholars, scientists and anyone interested in the machine vision principles employed in industry trends, development and prospects.

This book comprises 14 chapters. It was impossible to include all current aspects of the research in the targeted area. The book, however, presents a useful tool for several different methodologies to be applied to industry applications using machine vision. Each chapter reflects a various application field and methodology.

Chapter 1 presents challenges in Industry 4.0 for Machine Vision. This chapter aims at mapping the current state of industrial statistics analysis, identifying major problems that need to be tackled, addressing whether searching is on track to meet these challenges. Using these advanced technologies, however, not only offers manufacturers the foregoing advantages but also causes challenges such as handling large and heterogeneous data created by sensors and networked machines. This chapter is intended for readers wishing to begin Industry 4.0 research with comprehensive information and to outline, arrange, compile issues and challenges.

Chapter 2 focuses on Robotics Internet of Things. The combination of automated specialists and Internet of Things prompts the idea of the Internet of Robotics Things, in which development in computerised frameworks is attracting additional opportunities in both modern and research fields, covering a few areas, for example, fabricating, horticulture, well-being, reconnaissance and so on. It also covers how Internet of Robotics Things advances and applications are likewise talked about to underline their impact in regular day to day existence, inciting the requirement for more investigation into future computerised uses.

Chapter 3 analyses the role of sensing techniques in precision agriculture. Precise application and management of agricultural inputs is a challenging task in present-day agriculture. The green revolution focuses on improving crop productivity by use of genetically modified seeds, agrochemicals and fertilisers. Due to the excessive dosage of agrochemicals and fertilisers, there are severe environmental concerns and many health problems observed. In recent years, the prime focus has shifted toward optimising the agriculture inputs without any adverse effect on crop yield. By applying agriculture inputs at the right amount, right time, right place during crop cultivation, there is a chance to reduce the input cost. Artificial intelligence, the Internet of Things (IoT) and autonomous vehicles have become a part of the agriculture sector. The present study provides detailed applications of different sensors in the field of precision agriculture, i.e., irrigation, fertilisation, pest control, greenhouse.

Chapter 4 analyses the deep learning techniques for Industrial Internet of Things. The modern developments in the Internet of Things (IoT) are producing an increase in the abundance of interconnected devices, allowing a variety of elegant applications. The huge amount of IoT devices has produced a bulky capability of information that needs smart data study and processing methods. DL algorithms, when amalgamated in the Industrial Internet of Things (IIoT), can facilitate a mixture of applications for example, elegant assembling, effective manufacturing, efficient networking and accident identifying and prevention of errors. Manufacturers can make use of machine vision to achieve better safeguarding processes and facilitate them to construct real-time smart judgments derived from data. The main aim of this chapter is to focus on data acquisition, integration and predictive control in the industry using machine vision techniques for real-time collection and analysis of thousands of measurement points, early detection of equipment failures and process disturbances, detection of new types of anomalies as they occur for the first time, as well as Web-based UI for handling alerts.

Chapter 5 presents the study of machine vision for missing person location and identifier. Over the years, the number of missing people reports has increased. There are several reasons, some people may be hiding because they have committed serious crimes and others may have been abducted or even run away from their homes due to different social problems. This chapter proposes the system for the identification of missing people using biometric technologies and machine vision. This type of technology allows a system creation that contributes to the security strategy; as it facilitates the search for missing people and can directly replace all the disadvantages of using other conventional social media.

Chapter 6 presents a case study of inclusion of impaired people in industry. There are severe limitations for people who cannot perceive the correct instructions in a manufacturing process, resulting in accidents and injuries. This chapter proposes an intelligent interface that sends the essential instructions of manufacturing processes related to an emergency stop, a warning pause in the process and the running instruction based on detection of signs with hands. This study contributes to developing an intelligent assistant that helps deaf-mute people transmit instructions and warnings in a manufacturing process. It uses the convolutional neural networks (CNNs) as classifiers in the proposed assistant, which can then transmit the identified instructions as text and audio to be heard by the work crew.

Chapter 7 focuses on the study of a deep learning approach to classify the causes of depression from Reddit posts. The posts of subreddit are analysed to classify the reasons behind the depressed content, into various categories. The deep neural network algorithms such as long short-term memory (LSTM) and convolutional neural networks, have been trained using word embeddings generated by different techniques such as GloVe, FastText and Word2Vec. Also, the accuracy obtained by testing these models is further analysed.

Chapter 8 presents the Psychiatric Chatbot for COVID-19 using machine learning. The Indian Psychiatry Society estimates that the number of mentally ill people has risen by nearly 20% during the pandemic. Lots of people have sought out mental support and many are suffering. To provide artificial intelligence support, the authors propose multilingual chatbot enabling to address this issue with the help of an interactive interface to the end user by psychological behavior pattern analysis. The proposed Co-BOT has well-designed architecture along with a structured knowledge engine which supports and provides a language-independent interface with extendible scope to add further languages. The system also calculates the probability of a user contracting COVID-19 using parameters including age, gender and symptoms like fever and fatigue. The system also has an interactive interface with a virtual Avatar in a rich environment supported by virtual reality technologies with voice and language support.

Chapter 9 analyses drug-drug interactions using machine learning techniques. Many patients that are administered more than one drug need to be monitored for increased or decreased effects of drugs. Drug-drug interaction (DDI) is a significant operation for patient safety and efficient management in healthcare. DDI characterisation is extremely important to avoid undesirable drug reactions. This chapter proposes a framework called improved classification model (ICM) for accurate predictions of DDI types based on input drug details.

Chapter 10 presents image processing-based fire detection using IoT devices. In a wide or an open warehouse structure, where there are numerous places that are susceptible to a fire, general smoke sensors are not applicable. This chapter proposes a camera which actively processes images in real time and sends a warning if it detects a fire. It is faster than the traditional sensors. It gives a 2D view inside the warehouse and hence allows assessment of the situation.

Chapter 11 presents a case study for use of machine vision in transportation. In the populated country, the train journey experience is getting worse. It is observed that rush occurs in some railway trains more than in other types of trains. This study focuses on building an application that can predict crowd density in mass transport. The main objective of this work is to help people get more information about current crowd density. This will enable citizens to be aware of the rush in public transport. This will help in deciding whether or not to take the next public train or continue waiting. This will positively impact those companies that provide transportation service.

Chapter 12 provides an analysis of machine learning techniques to predict wine quality. The industries use product quality certification to sell or promote their product. The assessment of wine quality is done by human experts, which is a time-consuming process and is very expensive. The quality of wine not only depends on the amount of alcohol but it also depends on different qualities; these attributes change with time and so the quality of wine is also refined. It is important to determine the quality of the wine and to classify it into different categories on the basis of a quality evaluation. Different machine learning methods are used in this study to predict the quality of wine. This study presents a comparative study of fundamental and technical analysis based on different parameters. Various approaches are compared based on methodologies, datasets and efficiency with the help of visualisation.

Chapter 13 presents the applications and future directions in Industry 4.0. Machine Vision using a comprehensive set of sophisticated technologies to develop capabilities in machines to make self-aware and intelligent to get the visual representation of the surroundings. Innovations in machine vision, deep learning and robotics are revolutionising production lines and supply chains. Machine vision opens new ways of automation and digitalisation in an Industry 4.0 environment and the integration of this with robotics empowers intelligent decision-making abilities in smart design and production. If in tune with smart machines then the system is able to learn from its environment and take corrective actions to optimise production.

Chapter 14 focuses on the integration of modern technologies. Robotics and automation have created paradigm changes in the worldwide manufacturing industry during the past industrial revolutions. The Fifth Industrial Revolution could accomplish our approach to production and development with expectations and radical changes. In line with the Industrial Revolution 4.0, improvement required for the manufacturing industry is to be further strengthened. Industry 4.0 instruments, such as the Internet, digitisation, blockchain, additive manufacturing, artificial intelligence, swarm integration, robots, innovations in energy, biotechnology, virtual and enhanced reality are likely to change. The objective of this is to analyse and comprehend the previous research on the industrial revolution. In addition, this study creates a conceptual framework for Industry 5.0. Finally, it discusses the initiative and implementation of an integrated Industry 5.0 model in the manufacturing industry.

Machine vision combined with improvements in IoT continues to drive the value of diverse applications into other more linked contexts where vast opportunities exist to improve operational efficiency and productivity. These technologies enable IIoT connections, platforms and analytics solutions that can enhance productivity and cost-effectiveness, optimise income and unleash new business models. Data from connected items and systems can enhance production and efficiency while reducing expenses.

Many unresolved challenges remain, for example, how to make products, services and operations safer and scalable. There are no end-to-end security solutions, and many gadgets rely on open software platforms which make them more vulnerable. However, the technologies used in industry along with machine vision allow users to establish quick, flexible, high-performance systems that connect a wide range of devices and deliver data to serve many end-user goals and applications.

Dr. Roshani Raut
Pimpri Chinchwad College of Engineering (PCCOE)
Savitribai Phule Pune University
Pune, India

Dr. Salah-ddine Krit
Ibn Zohr University
Ouarzazate, Morocco

Dr. Prasenjit Chatterjee
MCKV Institute of Engineering
Howrah, India

Acknowledgments

The editors wish to express their warm thanks and deep appreciation to those who provided input, support, constructive suggestions and comments and assisted in the editing and proofreading of this book.

This book would not have been possible without the valuable scholarly contributions of authors across the globe.

The editors avow the endless support and motivation from their family members and friends.

The editors are very grateful to all the members who served in the editorial and review process of the book.

Mere words cannot express the editors' deep gratitude to the entire CRC Press/ Taylor & Francis team for keeping faith and showing the right path to accomplish this high-level research book.

Finally, the editors use this opportunity to thank all the readers and expect that this book will continue to inspire and guide them in their future endeavors.

Editors

Dr. Roshani Raut obtained a PhD in Computer Science and Engineering and is currently working as Associate Professor in the Department of Information Technology and Associate Dean International Relations at Pimpri Chinchwad College of Engineering, Savitribai Phule Pune University, Pune, Maharashtra, India. She has received various awards at national and international levels for research, teaching and administration work. She is working as an author/editor for various upcoming books of IGI Global, CRC Press/Taylor & Francis and Scrivener Wiley. She has worked on various conference committees as a convener, TPC member and board member. She has availed research and workshop grants from BCUD, Pune University. She has presented more than 70 research communications in national and international conferences and journals. She has published 13 patents. She has guided many UG and PG students in the area of artificial intelligence, machine learning, data mining, deep learning, Internet of Things and so on.

Dr. Salah-ddine Krit is currently Associate Professor at the Polydisciplinary Faculty of Ouarzazate, Ibn Zohr University Agadir, Morocco. Dr. Krit is currently Director of Engineering Science and Energies Laboratory and Chief of the Department of Mathematics, Informatics and Management. Dr. Krit received PhD degrees in Software Engineering from Sidi Mohammed Ben Abdellah University, Fez, Morroco in 2004 and 2009, respectively. During 2002–2008, he worked as an engineer team leader in audio and power management integrated circuits (ICs) research, design, simulation and layout of analog and digital blocks dedicated for mobile phone and satellite communication systems using Cadence, Eldo, Orcad, VHDL-AMS technology. Dr. Krit has authored/coauthored over 130 journal articles, conference proceedings and book chapters published by IEEE, Elsevier, Springer, Taylor & Francis, IGI Global and Inderscience. His research interests include wireless sensor networks, network security, smart homes, smart cities, Internet of Things, business intelligence, big data, digital money, microelectronics and renewable energies.

 **Dr. Prasenjit Chatterjee** is currently Dean (Research and Consultancy) at MCKV Institute of Engineering, West Bengal, India. He has over 90 research papers in various international journals and peer-reviewed conferences. He has received numerous awards including Best Track Paper Award, Outstanding Reviewer Award, Best Paper Award, Outstanding Researcher Award and University Gold Medal. He has been the guest editor of several special issues in different Scopus and Emerging Sources Citation Index (Clarivate Analytics) indexed journals. He has authored and edited several books on decision-making approaches, supply chain and sustainability modeling. He is the Lead Series Editor of "Smart and Intelligent Computing in Engineering", Chapman & Hall/CRC Press; the Founder and Lead Series Editor of "Concise Introductions to AI and Data Science", Scrivener-Wiley; AAP Research Notes on Optimization and Decision Making Theories; and Frontiers of Mechanical and Industrial Engineering, Apple Academic Press, copublished with CRC Press/Taylor & Francis. Dr. Chatterjee is one of the developers of two multiple-criteria decision-making methods called Measurement of Alternatives and Ranking according to COmpromise Solution (MARCOS) and Ranking of Alternatives through Functional mapping of criterion subintervals into a Single Interval (RAFSI).

Contributors

K.N. Agrawal
ICAR–Central Institute of Agricultural
Engineering
Bhopal, Madhya Pradesh, India

Harun Bangali
Department of Computer Engineering
King Khalid University
Abha, Kingdom of Saudi Arabia

Jaiprakash Bhamu
Department of Mechanical
Engineering
Government Engineering College
Bikaner
Bikaner, Rajasthan, India

Nivedita Bhirud
Department of Computer Engineering
Vishwakarma Institute of Information
Technology
Pune, Maharashtra, India

Ankita Biswas
Department of Computer Science and
Engineering
University of Calcutta
Kolkata, India

Pradnya S. Borkar
Department of Computer Science and
Engineering
Jhulelal Institute of Technology
Nagpur, Maharashtra, India

Yogini Dilip Borole
G.H. Raisoni Institute of Engineering
and Technology
Savitribai Phule Pune University
Pune, Maharashtra, India

Prasenjit Chatterjee
MCKV Institute of Engineering
Howrah, West Bengal, India

Edgar Gonzalo Cossío Franco
Instituto de Información Estadística y
Geográfica de Jalisco
Zapopan, México

Iván Alberto Cruz García
Instituto Politécnico Nacional
Ciudad de México, México

Jyotirmoy Das
Indian Institute of Technology Kharagpur
Kharagpur, West Bengal, India

Namrata Das
Department of Computer Science
and Engineering
University of Calcutta
Kolkata, West Bengal, India

Souvik Das
Indian Institute of Technology Kharagpur
Kharagpur, West Bengal, India

Luis Eduardo de Lira Hernández
Universidad Politécnica de Aguascalientes
Aguascalientes, México

C.G. Dethe
UGC Staff Academic College
Nagpur, Maharashtra, India

R. Dhaya
Department of Computer Science
King Khalid University–Sarat Abidha
Campus
Kingdom of Saudi Arabia

Ahona Ghosh
Department of Computational Science
Brainware University
Kolkata, West Bengal, India

Ananya Ghosh
Department of Computer Science and
  Engineering
University of Calcutta
Kolkata, West Bengal, India

Nikita Ghosh
Department of Computer Science and
  Engineering
University of Calcutta
Kolkata, West Bengal, India

Mihir Gune
Department of Computer Engineering
Vishwakarma Institute of Information
  Technology
Pune, Maharashtra, India

Héctor Manuel Gutiérrez Zazueta
Instituto Tecnológico de Culiacán
Culiacán, México

Mohammad Tariq Hasan
School of Business and Economics
United International University (UIU)
Dhaka, Bangladesh

Mohammad Amzad Hossain
Charles Sturt University
Sydney, Australia

Yogesh Jadhav
Computer Science and Engineering
Amity School of Engineering and
  Technology
Amity University
Mumbai, Maharashtra, India

N.K. Jain
Artificial Intelligence Group
CDAC
Delhi, India

Priyanka Jain
Artificial Intelligence Group
CDAC
Delhi, India

R. Kanthavel
Department of Computer Engineering
King Khalid University
Abha, Kingdom of Saudi Arabia

Apurva Kirdatt
Department of Computer Engineering
Vishwakarma Institute of Information
  Technology
Pune, India

Nilesh Bhikaji Korade
Department of Computer Engineering
PCCOE&R
Ravet, Pune, Maharashtra, India

O.B. Krishna
Indian Institute of Technology Kharagpur
Kharagpur, West Bengal, India

Krishna Kumar
Department of Mathematics
MIT School of Engineering
MIT Art Design and Technology
  University
Pune, Maharashtra, India

Pramod Kumar
Department of Mechanical Engineering
Government Engineering College
  Bikaner
Bikaner, Rajasthan, India

S. Vinod Kumar
ICAR–Central Institute of Agricultural
  Engineering
Bhopal, Madhya Pradesh, India

J. Maiti
Indian Institute of Technology Kharagpur
Kharagpur, West Bengal, India

Mahadi Hasan Miraz
School of Technology Management and
  Logistics
Universiti Utara Malaysia
Sintok, Malaysia

Arion Mitra
Department of Computer Science and
  Engineering
University of Calcutta
Kolkata, West Bengal, India

Ramiro Aguilar Ordaz
Instituto Politécnico Nacional
Ciudad de México, México

Kevin Gálvez Parra
Instituto Tecnológico de Culiacán
Culiacán, México

Pramoda Patro
Department of Mathematics (H&S)
Koneru Lakshmaiah Education
  Foundation
Hyderabad, Telangana, India

Swapnil Pawar
Computer Engineering Department
K.J. Somaiya College of
  Engineering
Mumbai, India

Yotziri Paloma Pérez Rios
Tecnológico Nacional de México/ITS
  de Ciudad Hidalgo
Ciudad Hidalgo
Michoacán, México

Dheeraj Rane
Department of Computer Science and
  Engineering
Medicaps University
Indore, Madhya Pradesh, India

Hanumantha Rao Sama
Department of Management
  Studies
Vignan's Foundation for Science,
  Technology & Research (Deemed
  to be University)
Guntur, Andhra Pradesh, India

Roshani Raut
Department of Information
  Technology
Pimpri Chinchwad College of
  Engineering
Savitribai Phule Pune University
Pune, Maharashtra, India
and
Pimpri Chinchwad College of
  Engineering
PCCOE, SPPU
Pune, Maharashtra, India

Martín Montes Rivera
Universidad Politécnica de
  Aguascalientes
Aguascalientes, México

Shumi Sarkar
Department of Business Studies
University of Information Technology
  and Sciences (UITS)
Dhaka, Bangladesh

Dharmendra Singh
Department of Mechanical
  Engineering
Government Engineering College Bikaner
Bikaner, Rajasthan, India

Sanket Sonje
Department of Computer Engineering
Vishwakarma Institute of Information
  Technology
Pune, Maharashtra, India

Farhana Rahman Sumi
Department of Business Studies
University of Information Technology
  and Sciences (UITS)
Dhaka, Bangladesh

Debabrata Swain
Department of Computer Science and
  Engineering
Pandit Deendayal Energy University
Gandhinagar, Gujarat, India

Subhash Tatale
Department of Computer Engineering
Vishwakarma Institute of Information
  Technology
Pune, India

Reena Thakur
Department of Computer Science and
  Engineering
Jhulelal Institute of Technology
Nagpur, Maharashtra, India

K. Upendar
ICAR–Central Institute of Agricultural
  Engineering
Bhopal, Madhya Pradesh, India

Alberto Ochoa Zezzatti
Universidad Autónoma de Ciudad
  Juárez
Ciudad Juárez, México

1 Challenges in Industry 4.0 for Machine Vision

A Conceptual Framework, a Review and Numerous Case Studies

Reena Thakur, Pradnya S. Borkar, Dheeraj Rane, Roshani Raut and Prasenjit Chatterjee

CONTENTS

DOI: 10.1201/9781003122401-1

1.1   INTRODUCTION

Industry 4.0 is commonly related to the development of deep learning, cloud computing and big data (BD), and, in particular, to a deep increase in cyber-physical use.
Systems such as sensors are capable of gathering data for manufacturers with producers to track and identify elements and subassemblies. These types of data gathering
approaches allow devices to unconventionally swap information as well as communicate and manage with each other independently, enabling operations that are more
automation-driven. The advent of Industry 4.0 will happen gradually and decisively
over a period of time, as with a transformation from previous "revolutions."

The fourth-generation industry started as an inventiveness of the "High Tech
Trends–2020 Plan" from the German administration to improve the development of the
country. Hundreds of millions of dollars have been spent by the German government
to support research through academia, industry and government. Although German
companies and their neighboring ones have earned the earliest, strongest interest in
Industry 4.0 innovations, among businesses all over the world, emerging ideas are now
growing. Generational growth from Industry 1.0 to Industry 4.0 is shown in Table 1.1.

1.2   INDUSTRY FOURTH-GENERATION TECHNOLOGIES

Several innovations have been working together to understand the utilisation of the
4.0 revolution in development as shown in the Figure 1.1.

1.2.1   THE INDUSTRIAL IoT (IIoT)

The Industrial Internet of Things (IIoT) exists to help data, computers and individuals in
the manufacturing world interconnect and collaborate. Essentially, it requires IoT – all
seamlessly connected and interfacing sensors, machines and data – and applies it to outputs. In the IIoT, every element of the production process can be related, and to improve
efficiencies in the production operation, the information it produces can be utilised.

1.2.2   AUTOMATION

The ultimate objective of a linked factory is to maximise productivity, thereby maximising profit. Automation can be integrated into some that includes much of the
production processes in an effort to accomplish this. Communication and intercon-
nectivity that occur in an integrated Industry 4.0 facility make automation possible
through robotics or artificial intelligence (AI).

1.2.3   AI—ARTIFICIAL INTELLIGENCE

A smart factory allowed by Industry 4.0 is basically a prerequisite for AI and its
subset of machine learning (ML). Manual processing is the whole idea surrounding

TABLE 1.1
Industrial Revolutions

Industrial Revolutions	Year	Major Features
Industry 1.0	Around 1760	1. The First Industrial Revolution was the transition to somewhat new processing methods using water and steam. 2. It has been extremely beneficial in terms of manufacturing a larger number of different goods, which creates a good standard of living.
Industry 2.0	Period between the 1760s and around 1840	1. During this time, new technological systems have been introduced, in particular, superior electrical technology, which has allowed for even greater production and more sophisticated machines.
Industry 3.0	Around 1970	1. It involved the use of electronics and IT (information technology) for further automation in production. Manufacturing and automation have made considerable progress thanks to Internet access, connectivity and renewable energy. 2. This revolution focused on automation onto the assembly line to carry out human tasks that mostly use programmable logic controllers (PLCs).
Industry 4.0	Around 2011	1. The Fourth Industrial Revolution is the era of smart computers, storage networks and manufacturing facilities that can independently share information, cause behavior and monitor each other without human intervention. 2. The main elements of Industry 4.0 shall include: • Cyber-physical system—a mechanical mechanism run by computer-based algorithms. • Internet of Things (IoT)—a collection of interconnected machines with devices through the Internet. • Cloud computing off-site network storage and data backup. • Cognitive computing—technical platforms that employ artificial intelligence.
Industry 5.0	Around 2019	1. Industry 5.0 is a revolution in which man and machine reunite and find ways of working together to improve the means and efficiency of production. The Fifth Revolution could already be underway among companies that have just adopted the principles of Industry 4.0. 2. It tries to produce human-centric solutions.

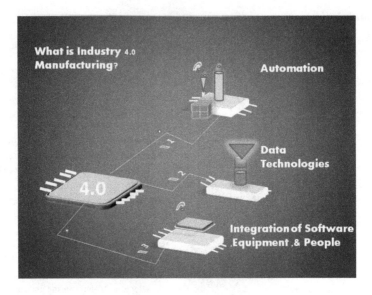

FIGURE 1.1 Industry 4.0 technologies.

this modern industrial revolution, and AI is the key method to be used in its place. To optimise equipment, reprogram workflows and generally recognise all changes which may be made to determine efficiencies as well as eventually sales, AI can expedite the data produced through a linked factory.

1.2.4 BIG DATA (BD) AND ANALYTICS

Since the entire production work is tracked and data are generated, tons of data are searched. Applications of BD analytics, however, use the ML and AI to process data rapidly and provide the information to decision makers to make changes in production operation as a whole.

1.2.5 THE CLOUD

Manufacturing companies do not want or choose to use much space to physically store vast volumes of data generated in an Industry 4.0 process. In a connected factory and a primary machine, it is the reason which makes cloud storage and computing an essential component. Cloud use also enables, at lightning speed, the only source of fact and sharing of data through the business. Finally, cloud storage also allows all data and computer operating systems to be remotely accessed and tracked, providing great visibility.

1.2.6 CYBERSECURITY

In Industry 4.0, since every touch point in the production activity is linked and digitised, an additional need for robust cybersecurity is required. It helps to secure industrial equipment, the clouds, data analytics, computer systems and every other IoT-connected device.

1.2.7 SIMULATIONS

The potential to predict performance is among the key manipulators in the context of Industry 4.0 and development. Prior to the actual digitisation of a factory, it was somewhat guesswork that often proves to be imperfect to turn over a manufactured goods line and maximise its speediness and production. Manufacturing facilities will enhance machinery with their links with highly emerging simulation models powered by IoT automation and analytics.

Industry 4.0 has a huge, nearly revolutionary impact on the manufacturing sector (Reischauer, 2018) and is therefore the topic of substantial debate (Liao et al., 2017). Industry 4.0 means incorporating, through steps associated, human actors, physical artifacts, intelligent devices, processes and product lines (Schumacher et al., 2016). Industry 4.0 contains a variety of innovations, from intelligent devices to IoT, which are used by industrialists to alter their approaches and processes (Schneider, 2018). The technological side of Industry 4.0 is preferred by the majority of scholars and specialists, while the organisational and managerial factors are still in their early stages (Horvath et al., 2019). In fact, few theories deal with the strategic dimensions of Industry 4.0 and concentrate on the impact of organisation of particular technologies of Industry 4.0. Managers therefore lack a detailed image of the innovations, challenges and advantages of this innovation wave. The fourth generation industry is also considered to be a package of emerging innovations to increase the performance of conventional production routines (Muller et al., 2018). In this context, it is important to undertake a comprehensive investigation of understanding both technological and organisational aspects of Industry 4.0 innovations, advantages and challenges. This chapter executes such a detailed view in two steps. Next, this chapter explains most significant technologies, challenges and benefits of Industry 4.0. The barriers and benefits are listed using a SWOT system. In this chapter, the term "E-SWOT" is being used to relate to both facilitators and the analysis of SWOT components' analysis. Several case studies in the next section explore the components of Industry 4.0 ESWOT with employees in different companies and areas (Li et al., 2016). In this step, a theoretical synthesis is carried out between the facilitators and the SWOT elements through an integrative analysis (Kiel et al., 2017a,b). A conceptual framework is built on the findings of these two phases to explain the "revolutions and evolutions" in Industry 4.0, and unique profits presently experienced through companies and the issues that stay behind. Among the furthermost significant contributions of this methodology, it discusses all the important enablers of Industry 4.0 simultaneously (Armando et al., 2020). In addition, the report is among the few to examine the management challenges involved in Industry 4.0. It needs to take the advantage of the analysis of companies of various sizes, industries and levels of services to compare Industry 4.0 enablers and SWOT components in this regard.

As shown in Figure 1.2, IoT supports various smart environments, and out of these smart factories, Industry 4.0 is one of the challenging fields in which the challenges and issues need to be taken care of in the various key points such as:

- Factory of future
- Machine-to-machine communication
- ML

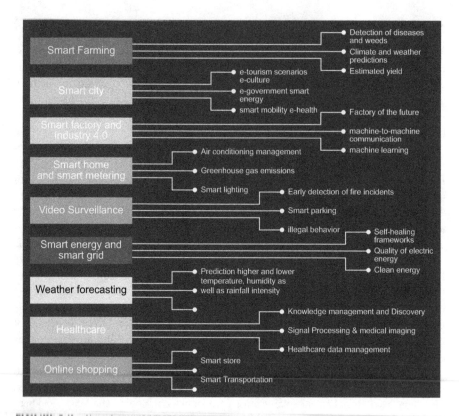

FIGURE 1.2   Smart environment with the help of IoT.

1.2.8   Conceptual Framework of Industry 4.0

We are accompanied by a number of devices which we call the IoT. From our mobile phones at home, we can turn on the heat. Pegs will tell us when it's time to bring the wash in, so it won't get dirty. Cars can alert drivers of imminent dangers coming so we can prevent an accident. Many of the "things" have been rendered under "Industry 4.0."

1.2.8.1   A Short Overview of Revolutions in Industry

We moved from manufacturing what we used in fields or workshops in the 19th century to large-scale production in factories. We have collected steam power and created engines that have led to the mechanisation of mass. Here is what we call the First Industrial Revolution, Industry 1.0.0. Electricity has moved everyone from the 1850s to World War I. Industry 2.0 was the Second Industrial Revolution. From the 1950s to the 1970s, knowledge became the next movement. As a result, computers made us all comfortable; everything started to become digital, and automation and robotics advanced into production components. This takes us to today, IoT and Industry 4.0. In places that are too dangerous for us to travel to, or with activities that are too difficult, computers are beginning to help individuals make decisions and work for us.

1.2.8.1.1 The Four Components of Industry 4.0

Four combined ingredients bring Industry 4.0 to us—(1) instrumented, (2) inter-linked, (3) inclusive and (4) intelligent which are described in the following.

1.2.8.1.1.1 Component—Instrumented

Data cover everyone. The appliances we bring about produce knowledge; information is generated by the cars we drive. Information is provided by the exercise and sleep monitors we wear. Data will be created by more and more of the goods and services with which we communicate. With all our eminent updates, videos, tweets, images and more, which are on top of the data, we create ourselves. With these, information is overflowing, and is getting quicker as well. As ARM (Advanced RISC Machines) as well as Texas Instruments found, they operate with IBM, the speed at which getting data from chips to analytics increases. Until it is moved, it can be analysed, all which we get the information is no good. That could be where it originates, in a cloud, or both, like the Blue Mix of IBM—all that information need to go somewhere safe.

1.2.8.1.1.2 Interlinked

We are linked constantly. After waking up, we look for our mobile phones first. Siri, the GUI (graphical user interface) you can connect to, was the latest OS update from Apple; Amazon's Echo plus Google's Home both make it much easier to interact with our smart devices and do something for us, such as talking back. These are clouds that manage and store our data, and "fog computing" does what clouds do, but on the "network edge" on the field, such as IBM and Cisco.

Platforms, such as the Watson IoT Platform, also exist to help process information. Even if you are associated with the "Hitch Hiker's Guide–Galaxy," these platforms provide a place for data collection, converting it into a standard template that can work with the brain's computing, and transferring insights and results back. You can use this simple language called MQTT if you've ever used the "Messenger" app on Facebook.

1.2.8.1.1.3 Inclusive

Did you have an alarm acknowledging that the weather was going to be a mess? Poor weather is going to obstruct the traffic. Your drive takes longer as a result of the busy traffic, so your alarm wakes you up a little faster, so you're not at work late. This seems to be an example of three data sets that can predict some need for you in conjunction with a computer and change your environment to handle it.

Knowledge they generate may also become a new revenue source for certain businesses. Catch the modest umbrella. The forces of Mary Poppins were unique, and you could think of linked umbrellas as well. A fabric maker will like to understand where umbrellas have been used to provide durable fabric to shops that could sell very well display advertising with offers to promote customers to purchase it. It can be beneficial for an estate professional to understand precisely how often people are going to operate with raincoats to arrange the mop floors of cleaning operators to avoid them from mixing up and adjusting the temperature to make it comfortable for rainy employees. A cab company would really like to know how raincoats were arranged to adjust the location of its fleet to take account of travelers who did not want to stay in the rain. Awareness of background gives insight and can improve sales.

These alliances will become even more meaningful and, like never before, new relationships will develop within businesses and through industries. Having a place to strengthen and develop industrial relations, accompanied by the opportunity to promote them, such as the IoT HQ of IBM in Munich, would see more alliances being formed that will benefit customers.

1.2.8.1.1.4 Intelligent   Knowledge is fourth, but not least. With the utilisation of cloud computing and the complexities of science, more informed decisions can be taken using all the "big data." These options, like IBM's Watson, come with a degree of trust. Cognitive computing, ML, predictive analytics and AI are learnt through the knowledge provided around us by the rising number of connected sensors. The world is now entering into a new Industrial Revolution with these four elements in place. Industry 4.0 is revolutionising the way items flow through the market.

For example:

Car services: the new definition of vehicle ownership is evolving. With the help of your phone, a person can use another car to open the door then disburse the payment for the ride; the person can phone a car through an app and rent a car within an hour. The days of buying or renting a vehicle, worrying about insuring, parking and financing are over for you now. Alternatively, your car can prepare your journey, considering the effects that traffic, keeping you safe in your trip, knowing when to pause, even knowing when there is a break in traffic.

Instead of getting your car serviced, you can now employ a voice-enabled edge to upgrade your car's software. It may become as easy to upgrade your car as it is to upgrade your computer. For manufacturers, designing innovative features which the consumer wants and ensuring quality products will be crucial. Companies are now using this comprehensive industrial engineering system to achieve their entire engineering processes for designing, creating, evaluating, implementing and controlling their high-end systems.

The sensors of today will modernise old properties. Steering the designs of the tomorrow is the way we use resources now. With the massive remote control that still ends up on the side of the sofa, replaced with a tiny control that only has what you really want, or no physical control, you just tell your TV what you need it to do or rock at it.

Product design insight comes from the way we use products today. In the very near future, the data we produce steer the goods we will purchase. It will, indeed, guide not only product design but also move manufacturers into new areas. Michelin is known for its tires. Today, they can manage fleets better with connected tire data; guide drivers on how to save fuel and stay safe through changes in the way they drive; become instructors, teaching other companies how to travel the road more efficiently. Michelin moves from a tire manufacturer to a driving instructor.

1.2.8.2   Industry 4.0—Promise and Advantages

1. Competences in Industry 4.0 can convert raw facts into actionable knowledge and insights that can drive immediate, measurable changes in results. It will do this by analysing "big data" collected during the development

process and allowing cloud computing and deep learning techniques to discover patterns for process enhancement. Data can be obtained in many situations today, but they remain isolated and solitary. These data can be proactively exchanged and used in the future as well as driven by data-rich technology such as machine vision.

2. The communication networks of manufacturing systems will search for feedback through the market and then use information to obtain accurate output parameters. Such systems of a pharmaceutical industry notify higher than anticipated incidents of allergies discovered in a specific area, which trigger manufacturing lines to expand supply and position additional raw material orders.

3. Equipment for the automation line can include growing levels of self-learning, self-control and self-optimisation capabilities to complete complex tasks quickly and accurately with minimal interference. In combination with operator guidance, this form of capability is available today in some revolutionary devices. However, with the advent of Industry 4.0, it will become even more autonomous.

Since robots, inputs of raw materials, machine vision systems and other manufacturing line components are capable of communicating directly with each other, greater versatility can be enjoyed by manufacturers. The advantages can comprise the skill to manufacture a broader series of parts on a single line, or the more cost-effective manufacture of smaller batches of particular components, including personalised goods.

1.2.8.3   Industry 4.0—Challenges

1. **Capital:** This kind of transition would require huge investment, which is not inexpensive, in new technologies like AI, IoT or robotics. At C-level, the decision to make such transformations would have to be made. Nevertheless, the risks have to be calculated and to be taken seriously. In addition, such technologies will require tremendous capital that alienates small companies and could lose a company's market share in the future.

2. **Security:** IT protection risk is the most difficult part of introducing the Industry 4.0 platform or techniques. Data leaks and security breaches may be caused by online integration. Cyber fraud should also be taken into account because, in this case, the issue is not individual, but the money will and possibly would cost the suppliers, and it could also harm their reputation. Adequate research on "defense" is therefore critical.

3. **Privacy:** Privacy of data is not only the consumer's issue, but also the supplier prime concern. Producers should collect and analyse the data in the interlinked industry. This could appear to the buyer a risk to anyone's security. Not only is this unique to consumers, but it would also have to aim for a much more transparent atmosphere for big or small corporations that have not exchanged their data in the past. It will be a major challenge for both the parties to bridge the gap between producer and customer.

1.2.9 MACHINE VISION

Machine vision is a set of technologies that gives machines greater awareness of their environments. It promotes the understanding of higher order images and decision-making based on that knowledge.

An integral aspect of the automation system is machine vision. No other part of the production line gathers more information or is more useful in product assessment and impact finding, as well as in data collection to guide the process and maximise the efficiency of robots and other equipment. Unlike simple sensors, in an Industry 4.0 environment, vision sensors produce vast volumes of image data, intensifying their usefulness. The large amount of data available across vision equipment could be used to identify and mark faulty products, recognise their limitations and allow rapid and efficient action as the data analytics capabilities advance in the Industry 4.0 factory.

1.2.9.1 Historical Perspectives

Introduction: in recent years, due to a higher degree of digitalisation, industrial production processes have been changed, leading to smart, wired and decentralised production. This new stage of organisation is also known as "Industry 4.0" or "The Fourth Industrial Revolution" (Kagermann et al., 2013; Hermann et al., 2016). The key principle of Industry 4.0 is the use of modern technology as a medium that deeply blends market and engineering processes to operate production in a flexible, productive and sustainable manner, with consistently high quality and low cost (Wang et al., 2016).

In general, the implementation guidelines for Industry 4.0 are driven by government initiatives, such as Sweden—"Smart Industry" and "Produktion 2030" (Ministry of Enterprise and Innovation, 2016; Teknikforetagen, 2017); Japan—"Super Smart Society" (Keqiang, 2015); European Commission—"Factories of the Future" and "Horizon 2020" (European Commission, 2016); France—"La Nouvelle France Industrielle" (Foresight, 2013); the United States—"Advanced Manufacturing Partnership" (President's Council of Advisors on Science and Technology, 2014); the United Kingdom—"Future of Manufacturing" (Foresight, 2013); Germany—"High Tech Strategy 2020" and "Industry 4.0."

Manufacturing industries need to deliver a better return on investment while reducing the environmental impact. They will need to create a beautiful office for individuals to focus on teamwork, knowledge and skills growth. The Industrial Revolution 4.0 is, according to the author (Ngjeqari, 2016; Beier et al., 2018), an enabler of sustainable development; however, the implementation of technology transformation and conservation appears underdeveloped. Recently, some studies have identified several interconnections between Industry 4.0 technologies and operational processes. For example, the authors in De Man et al. (2017) discuss the effect of Industry 4.0 on viable business models, Kamble et al. (2019) discuss the effect of Industry 4.0 on sustainable manufacturing practices, Ghobakhloo (2019) introduces the influence of dependency of the determinants for the implementation of Industry 4.0 in the sense of sustainability.

1.3 INDUSTRY 4.0—AN ENABLER

This section will describe the various enablers with their description.

Table 1.2 shows the different groups for Industry 4.0 enablers that categorise technology sharing, data analytics, network technologies, cybersecurity technologies for manufacturing, employees and equipment.

1.4 CASE STUDIES

This chapter comprises a few case studies.

1.4.1 CYBERSECURITY IN IoT

The security trends in the concepts of Industry 4.0 and enabling components:

1. *CPSs*: These (cyber-physical systems) work on TCP/IP (Transmission Control Protocol/Internet Protocol), so the most commonly known threats are man-in-the-middle attacks and IP spoofing
2. *IoT*: Trust, secure middleware, privacy, access control, authentication and confidentiality
3. *BD*: Secure computations in a cryptographically enforced access control, transaction logs, secure data storage, distributed setting which are attack areas

TABLE 1.2
List of Industry 4.0 Enablers

Name of Enablers	Details
Technologies for Production Line	Technologies include equipment for interactive and cooperative robots, commutative, additive manufacturing like 3D printing installed on the technology production line.
Technologies for Smart Workers	Wearable sensors like intelligent glasses which connect the virtual and physical worlds are included in this class.
Smart Equipment Technologies	These technologies include sensors and monitors for factory machinery control and identification.
Computing Technologies	Technologies that are needed to model and simulate models of the production system involved in this class.
Technologies for Sharing	Parts of this class are the techniques that enable resource sharing.
Technologies of Smart Goods	This class is similar to smart enablers, but they are built on complete goods.
Data Analytics Technologies	The technologies needed to work and analyse huge volumes of the data are included in this class.
Technologies for Networks	Technologies that enable air communication among physical objects are included in this class.

4. *Cloud computing*: Interface hacking, denial of service, data loss, advanced persistent threats, malicious insiders and account hijacking
5. *Colorant manufacturing*: Confidentiality, integrity and availability
6. *Smart sensors*: Large attack areas due to proliferation of the smart factor
7. *ML*: Attackers can use adversarial samples to extract sensitive information embedded in the model or force the model to mispredict it, which leads convincing consumers to avoid the model because of its poor performance
8. *Robots*: Large attack areas for robots make them an easier target
9. *Augmented reality (AR)*: Private information from users or confidential data from an organisation from various sources, such as computer monitors or audible and visible moving sections of machines

There are numerous challenges to extensive implementations of the IoT in the field (Li et al., 2013; Fukuyama, 2018; Manglani, 2018). Avaneviciene et al. (2019) categorised these challenges into three broad categories:

1. Strategic
2. Tactical
3. Operational

This analysis focuses on the strategic point of view, which entails IoT problems of social responsibility, i.e. cybersecurity. The difficulties listed in Table 1.1 are not mutually exclusive. Poor or uncoordinated policies, for instance, may lead to service protection concerns, which could lead to privacy problems and loss of profitability or lack of appropriate skills/knowledge and could exacerbate interoperability and integration problems, leading again to a loss of profitability. The argument is that the boundaries are exchangeable, and problems may be grouped in many ways, although they may remain the same in general. The cost problem in the IoT could easily be explained when covering all the other challenges, but because of the understanding of the IoT as a service phenomenon, it was preferred to be explained through cybersecurity.

1.4.1.1 A Comparative Analysis of the IBM X-Force Cybersecurity

Table 1.3 shows the top 10 targeted industries. Kemper et al. (2016) highlighted the difficulties of implementing Industry 4.0, especially in the fields of fabric machinery and manufacturing induced by sub-sectorial inconsistencies (roving or sliver, yarn, cloth, clothing, finishing) and the disparity in the level of mass, duration and manufacturing in each sector. In 2017, the Association of Italian Manufacturers of Textile Machinery (ACIMIT) conducted a survey on the feasibility of applying Industry 4.0 in three sectors: design, manufacturing and machine maintenance (ACIMIT, 2017), while Italian manufacturers of textile machinery replied to a number of questions and ranked the feasibility of applying Industry 4.0 to machine maintenance as the highest. Table 1.4 shows the attack sources by industry.

TABLE 1.3

Top 10 Targeted Industries

Sector	2019 Rank	2018 Rank	Change
Financial services	1	1	–
Retail	2	4	2
Transportation	3	2	−1
Media	4	6	2
Professional services	5	3	−2
Government	6	7	1
Education	7	9	2
Manufacturing	8	5	−3
Energy	9	10	1
Health line	10	8	−2

1.4.2 ISSUES IN TEXTILE FACTORIES

The Rina Consulting Group's fascinating SWOT review for ACIMIT indicated a lack of expertise and training along with low interoperable standards that would adversely affect the adoption of Industry 4.0. The need for assistive policies and government funding to support IoT implementation in the textile industry was another significant point made. In Perez et al. (2016), the authors addressed that the challenges and essential technologies needed to allow a smart textile factory and the four critical requirements for textile manufacturing were information carriers that should become central, warp beams and fabric (Gloy et al., 2015). Assigning knowledge to commodity carriers leads to autonomous textile process chains at the heart of the world of CPSs, with human-machine interaction (M2H).

TABLE 1.4

Attack Sources by Industry

Sr. No.	Industry Sector	% Malicious Insider	% Inadvertent Actor	% Outsiders
1	Financial sectors	5	53	42
2	Communications and information sectors	1	3	96
3	Manufacturing sector	4	5	91
4	Retail sector	2	7	91
5	Healthcare sector	25	46	29

1.4.2.1   Challenges for Handling Weaving Machines

A number of technical research studies were published at RWTH Aachen University on the CPS weaving machine project listed by the authors in Kerpen et al. (2016), who founded the research group SozioTex, which developed augmented human-machine interaction (M2H) and communication aimed at helping as well as improving the aging population. This task was accomplished by using smart glasses, mobile applications and an optimisation algorithm. Unskilled labor, wearing these smart glasses, would be able to learn about machine parts and operations faster as well as being able to detect broken yarn feeds in the circular knitting machines. In Kerpen et al. (2016), the authors explored the installation of a digital environment in the weaving mill based on AR; contrary to what is believed to be fundamental to automation, his research took human-machine interaction as a cornerstone involving the sharing of bidirectional information. By removing monotonous and repetitive activities and supporting them, the digitalisation of production improves the skills of the workforce. In Saggiomo et al. (2016), based on a multi-objective input algorithm, the authors optimised warp stress, energy consumption and fabric quality to create a weaving machine as a CPS. The authors used smart bobbins with integrated RFID (radio frequency identification) in Saggiomo et al. (2015) and Simonis et al. (2016), which could be read on the weaving machine to record inhomogeneities and spliced components, the location of the defects and the direction of the defects and exact running time of bobbin-material utilisation in the creel. A similar approach can be used to build carding/drawing/roving machines such as CPS, using AR to detect yarn breakage as well as to increase the performance, training and versatility of work in yarn-spinning plants. Automated sewing machines and fabric-defect-detection systems are other major development segments of IoT applications in the textile industry, from a process perspective. While they have the greatest ability to exploit the application of ML and AI, these solutions actually support manual operators. For example, the sensing and monitoring of sewing machine parameters for a smart manufacturing process was published in Lee et al. (2017); the automated sewing machine faces the challenge of fabric control to achieve constant seam width, maintain proper tension of the fabric and avoid buckling of the fabric. The author approached fabric control for automatic sewing in Winck et al. (2009), which claims to solve these issues. Currently, SoftWear Automation, Inc., located in Atlanta, uses this fabric control technique which has recently been delivered.

1.4.2.1.1   Challenges for Sewing Rolled-Goods Fabrics in the Textile Industry

As per the report of textile progress of 247 sewbots to Tianyuan, China, 800,000 T-shirts are made a day for Adidas at a rate of 22 seconds per T-shirt at a cost of 33 cents (Borneman et al., 2017). The author provides an interesting algorithm to detect broken needle lines using smart visual sensors.

Another example reviewed was the automated sewing of rolled fabric goods at Yeh Group Inc., a Thailand-based performance fabric manufacturer for global sports brands. This work automated the butt-sewing of two rolled-goods fabrics across their width before feeding them into a dyeing machine. The feed dog mechanism was not changed, as the author discussed since the sewing is temporary, unlike the case for garments. The fabric feed was automated using a 3D gantry system using Festo linear drives (providing movements in X, Y and Z directions) and smart placement of intelligent sensors for drive position, drive speed, drive feed, machine position and speed. This work is still in the development phase, but in future, this work will include processing information using a local programmable logic controller to communicate data to the central server on company premises over ethernet to "create net-based" application web pages (using open-source Microsoft Visual Basic) for operator and manager interaction. In the Yeh Group, work on fabric defect detection was carried out using Omron's vision sensors (FZ3 Series) according to Adidas 4-point defect system, which follows ASTM 3903. An algorithm was developed to improve auto-shutter speed change via CIELAB and RGB colorimetry techniques overcoming some common industrial problems for vision sensors like the inability of the sensor to differentiate between noninherent and natural creases; future work includes ML where each new defect detected by the sensor will be set aside and tackled with the new technique. Texdata International reviewed the use of digital twins in manufacturing. They defined digital twins as "a virtual model of, for example, a process, product or service which combines the real world with the virtual world." Examples from the textile industry included the digitalised production of fiber composite components by DLR(German: Deutsches Zentrum für Luft- und Raumfahrt e.V., literally German Center for Air- and Space-flight). DLR fully automated the process chain using resin-transfer molding at the DLR site in Stade, Germany, by adding a digital twin to a CPS to monitor the real-life production through virtual means. The aerospace industry needs highly reliable products with an extremely high safety factor, which warrants the extra cost of having digital twins even for associated textile products. Another example is Adidas, which has collaborated with Siemens to create a digital twin for the Adidas Speed factory, and by doing that it is claimed shortening time-to-market, increasing flexibility and boosting product quality and efficiency. Research for the digital twin of the Adidas Speed factory is also being carried out at the Institute for Factory Production and Automation Systems in Erlangen. An important observation was the requirement of high-performance software for creating digital twins such as Siemens MindSphere, SAP S/4 HANA (SAP Leonardo IoT), PTC's THINGWORX or Ansys Twin Builder along with product offerings from IBM, Microsoft, C3 IoT, Software AG, Hitachi, GE Digital, Atos, Oracle, Bosch, AWS and Schneider Electric.

1.4.3 DEFECT DETECTION

The incidence of defects raises expenses and worsens production processes. In different industries, systems for defect detection exist. Extensive research has been performed to analyse the effects (Hayek et al., 2001), improve planning (Chiu et al., 2007, 2014) and reduce the costs (Taleizadeh et al., 2010) of accidents involving defects. While some methods only apply to unique circumstances, while others effort in a broad range of domains. Computer vision algorithms (Viharos et al., 2016), artificial neural networks (ANN) and Gabor filters are usually used for physical defect detection solutions. Defect detection methods, such as multilayer perceptron (MLP), Bayes classifier and learning vector quantisation (LVQ), typically use image-processing methods along with analytical, numeric or stochastic algorithms. A non-vision approach with the more standard theory by the author automatically learns from output data using probabilistic deterministic timed automation to identify certain defects (PDTA).

Although the method is more specific and could respond to changes, no details on how the model will alter or develop during runtime are included in the work.

1.5 REFLECTION OF INDUSTRY 4.0 IN BUSINESS

Table 1.4 illustrates how Business 4.0 is changing the organisation of business. The inputs, outputs, dynamics, market structure and production processes by which price added is produced and assumed are evolving. Nearly all manufacturing operations, from the input side, face a development of the skills required for operating and competing. Currently, AI, biotechnology and data science are the main inputs for agrobased development, and AI and BD are significant for tracking energy efficiency and innovation across all industries (OECD, 2017, 2018). Industry 4.0 is altering the skills necessary for successful results by employees and managers. For example, the business aspect of cognitive science is coupled with coding and programming skills to pick managers for sales of food items.

1.5.1 MAIN AREAS OF STRATEGY FOR SHAPING THE FUTURE

Many countries which are considered under the developed countries need to deal with an agenda which is dynamic one, which may go beyond the consensual necessary areas as shown in Figure 1.3. Much of the current debate focuses on learning the right future skills and accelerates the transition to the infrastructure which is digitally linked. Two priorities, both public and private, are important, which require investment. However, Table 1.5 depicts the important key areas which may play a

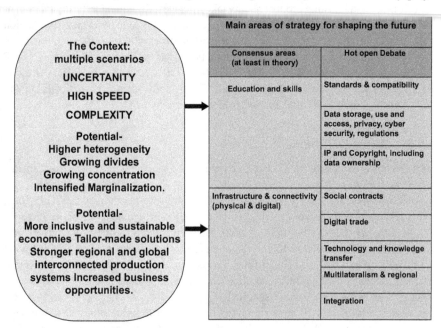

FIGURE 1.3   Main areas of strategy for shaping the future.

TABLE 1.5

Industry 4.0—How and What Is Changing?

What Is Changing		How?
Inputs	Capabilities and new raw materials	For electric vehicles, e.g. electric cars, input needed to be lighter. New, growing electronics content of traditional production outputs.
	Jobs	Cognitive science, data science, etc. The overall number of workers has been affected. Tasks will be redefined, as many roles will be assumed by digital technologies, not just routine ones. New specialists (such as big data analysts) will emerge. An estimated rise in entrepreneurial numbers.
Outputs	Interconnected and integrated devices personalised mass production from goods to services that are integrated (such as from vehicles to mobility services).	
Production and services delivery processes	Within firm organisation	Management, logistics and payment systems that are managed by digital and artificial intelligence.
	Between firms (B-B)	Digitally controlled logistics: off-site operations are controlled remotely. Supplies all associated information and with links to real-time information.
	Consumers-producers (C-B)	Digital platform is used to interact between consumer-business and business-consumer.
Market structure	Scenario:	
	1. Large companies which have little business expertise try to dominate all markets.	
	2. Digital public forum and scattered industrial ecosystems.	
Value generation and capture	For businesses, data become an asset as depending on the data, new ideas can be generated.	
	It will entail new modes of governance for data access, usage and governance.	

more passive role in developing countries. Out of these, digital standards, data and intellectual property and intangible property problems, as well as cybersecurity, privacy, taxation and digital commerce are not worth mentioning.

1.5.1.1 Industry 4.0—Outcomes

It has been found that Industry 4.0 has an international impact in terms of communication across continents and it is also important to consider the way it affects global economies. In the United Kingdom, Deloitte Analytics reported that, between 2001 and 2015, these emerging technologies helped to create 3.5 million new jobs.

These innovations are helpful to understand on a wide-reaching scale of innovative ideas; so the question is how do they impact the individual producer? Thus, there are five of the biggest and most meaningful benefits that manufacturers can expect from Industry 4.0.

1.5.1.1.1 Optimised Processes

All networking sensors, IoT, AI and so on, from Industry 4.0, serve one primary purpose: optimising production processes. Computerisation helps manufacturers to work faster, data analytics enrich administration to make data-driven efficiency maximisation decisions and quality control means less computer downtime, and monitoring systems provide real-time operational optimisation of yields.

What does a producer really mean by streamlined processes and maximised efficiency? This leads to sales growth and better customer support in the case of digital transformation and Industry 4.0. When manufacturers with sensor-monitored machines are able to get the most out of their production, they can really see the importance of the linked factory, all while providing personalised attention and fast service to customers through AI and field services.

1.5.1.1.2 Greater Asset Utilisation

The Fourth Industry provides better versatility in the industrial activity, resulting in better use of assets and hence a potential for sales increases. Autonomous mobile robots can manage small tasks like product transport, allowing professional human staff to perform more high-value tasks. Think about automation.

1.5.1.1.3 Higher Labor Productivity

They will focus more and perform more tasks during the day when workers feel more relaxed in the workplace. Worker security is one of the key benefits of IoT solutions for on-site floor sensor manufacturing and is constantly monitored by workers to ensure a healthy and stable working climate.

The Fourth Industry also expands many manufacturing workers' range of abilities. Employees are learning new skills that enhance organisational productivity and their skills as digital technologies come into play. Think of cobots (collaborative robots)-people and robots working together to optimise efficiencies and profits in production workflows.

1.5.1.1.4 Supply Chain and Inventory

IoT allows data analytics and sensors to provide producers with insight into the whole supply and manufacturing process. Combining these levels of visibility with ML and AI capabilities gives supply chain optimisation to be done in real time. Some may call it Supply Chain 4.0, characterised as the use of the IoT, the use of advanced BD analytics and the advanced robotics use, in supply chain management: the development of networks everywhere, the placement of sensors in everything, the analysis of everything and the automation of everything, to dramatically improve efficiency and customer satisfaction.

1.5.1.1.5 After-Sales Services

Remote monitoring, virtual reality and predictive analytics that are foundations of Industry 4.0 are again translated after development into the customer space.

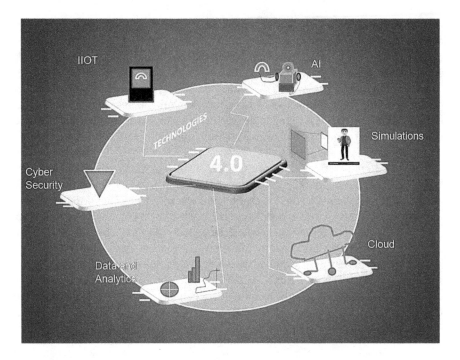

FIGURE 1.4 Business outcomes.

Although this does not have a direct impact on a producer, if they produce products that are capable of IoT communication, customer and field services can be dramatically enhanced as shown in Figure 1.4.

1.5.1.2 Revolutions Are Disruptive

The revolution in technology is innovative in nature, and Industry 4.0 cannot be excluded from this revolution. Much of the labor work, routine work or work in an unsafe environment and manual work will be replaced by sophisticated "thinking" robots. New tasks may emerge that involve new abilities. In the early days, the supply of such services could be too limited, restricting businesses transitioning to Industry 4.0 models. There is risk where there are data. There are hackers leading on from danger who can hack. As a result, it will be paramount to maintain stable networks and keep data protected. This is especially significant when humans take into account all the available devices we use and the associated "things" in which we communicate. Technology will continue developing rapidly.

1.6 FUTURE TRENDS

A significant shift in viewpoint is rendered by Business 5.0; the center in Society 5.0 emphasises individuals as the primary axis of the manufacturing field. These two fields of manufacturing and marketing accept that Society 5.0.0 is outside the focus of Industry 5.0. The services and products offered in Society 5.0 will be customised to the client's requirements. The purpose is to improve a convergence of technological development and human beings, with the primary objective of commending humans and machines with their activities, the use of cobots and robotics is a radical shift in routine, dangerous and risky work collaboration. In addition, intellectual development will be human activity, which means that it will be important to be skilled to be constructive in this model of society as shown in Figure 1.5.

It is expected that this new man-machine interaction approach will increase output and offer, and with personalised goods, both the worker and the final client become happier. Thus, it is found to be important to recognise that Industry 5.0 goes beyond a method of development and it strives to integrate a Society 5.0, thought and creating for humans and cobots.

The paradigm based on individuals is one of the major inconsistencies with previous generations. In the manufacturing process, process automation, the cobots advent and technological evolution enable individuals to learn new skills.

New strategies, hardware and software are needed in this innovative approach and must be integrated with cobots and high-tech human training for Society 5.0.0. To achieve intelligent education, it is necessary to shift from its traditional shape. Industry 4.0 is considered to be at the center of Industrial Revolution. It has robots and other technological foundations, while technology enhances or collaborates with human worms in Society 5.0.

It is important to highlight that by introducing human creativity and craftsmanship to the focus on the efficient process, technological innovations made in Industry 4.0 enhance

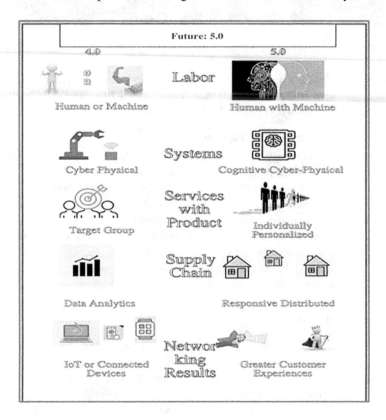

FIGURE 1.5 Future 5.0.

the efficacy and effectiveness of Society 5.0. The circular economy and environment as a goal are another essential aspect of this new society, although a different vision is expected.

It must also be remembered that the goods and services can be personalised in compliance with the actual market demand to prevent oversupply and not saleable inventories found in Industry 4.0; the primary concept is to generate what it needs for population, resulting in minimal over-cost which may reduce overproduction.

The vision is of a smart society in which smart industries, smart cities, autonomous vehicles, and so on, are incorporated, in addition to climate change action: eliminating all undesirable problems in the management structures and business world.

The ideal idea would be to have a more helpful society, with more cooperation between people, more trust among individuals, communities and nations, greater trust and credibility among customers, negligible corruption in public administration, equal distribution of resources; and businesses would form the basis of the market-based development process.

Society 5.0 is meant to be more environmentally sustainable and multicultural; at the same time, it is also intended to have a greater regulation of autonomy and assimilation with the society and world. The general expectation is a more sustainable world, which is needed where all impactful issues such as environmental, economic and social issues are related and tangled.

1.6.1 INDUSTRY 5.0

Robots and machines work together with people in Industry 5.0 as shown in Figure 1.6. With the help of BD and the IoT, robots help humans in a better way and get results

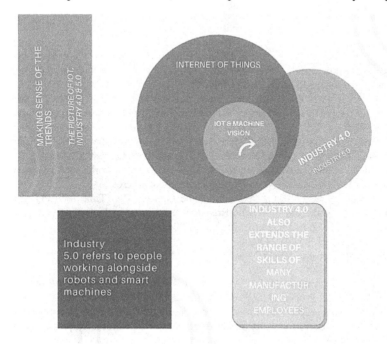

FIGURE 1.6 Relation Industry 5.0.

faster. This gives the personal touch in terms of human to Industry 4.0, which forms the pillars of automation that increases efficiency.

Robots have done risky, hazardous or physically challenging work like painting and welding in car factories. In manufacturing environments, robots have historically performed dangerous, monotonous or physically demanding work, such as welding and painting in car factories and carrying heavy loads in warehouses. Machines at the workplace are becoming smarter and better connected. Industry 5.0 focuses on collaborative operations by combining human intelligence with resources.

Most of the time, industrial robots can be operated separately from workers and behind safety measures. The combination of human and machine workers is found to open new opportunities in manufacturing companies. Manufacturers should smartly strategise various ways to integrate human and machine workers in a way to increase benefits that can be collected as the movement continues to develop.

1.7   CONCLUSION

The present vision and future of Industry 4.0 are discussed in this chapter. It also includes the history of Industrial Revolution. It finally shows what is Society 5.0 and a future vision of Industry 5.0 is expected to begin. The main aim is to focus on customer value that always comes before any technical solution to any problem. As we see, many times technology comes and goes but the quest for crucial value remains.

REFERENCES

ACIMIT, (2017). Available at: www.acimit.it/DOC/Executive17-Ing.pdf.
Armando, C., Nathan G., et al. (2020). "'Evolutions' and 'Revolutions' in Manufacturers' Implementation of Industry 4.0: A Literature Review, a Multiple Case Study, and a Conceptual Framework", ISSN: 0953-7287 (Print) 1366-5871 (Online) Journal homepage: https://www.tandfonline.com/loi/tppc20.
Avaneviciene, A., et al. (2019). "Individual Innovativeness of Different Generations in the Context of the Forthcoming Society 5.0 in Lithuania." The Engineering Economist 30 (2): 211–222.
Beier, G., Niehoff S., Xue B. (2018). "More Sustainability in Industry through Industrial Internet of Things?" Applied Sciences 8 (2): 219. doi:10.3390/app8020219.
Borneman, J., Textile W. (2017). Available at: https://www.textileworld.com/textileworld/features/2017/09/u-s-textiles-have-no-fear-automation-is-here/.
Chiu, S., Wang S., et al. (2007). "Determining the Optimal Run Time for EPQ Model with Scrap, Rework, and Stochastic Breakdowns." European Journal of Operational Research 180 (2): 664–676. doi:10.1016/j.ejor.2006.05.005.
Chiu, Y. S. P., Chang H. (2014). "Optimal Run Time for EPQ Model with Scrap, Rework and Stochastic Breakdowns: A Note." Economic Modelling 37: 143–148. doi:10.1016/j.econmod.2013.11.006.
De Man, J. C., Strandhagen J. (2017). "An Industry 4.0 Research Agenda for Sustainable Business Models." Procedia CIRP 63: 721–726. doi:10.1016/j.procir.2017.03.315.
European Commission. (2016). Factories of the Future PPP: Towards Competitive EU Manufacturing. European Commission. http://ec.europa.eu/research/press/2013/pdf/ppp/fof_factsheet.pdf.
Foresight. (2013). The Future of Manufacturing: A New Era of Opportunity and Challenge for the UK. London: Government Office for Science. doi:10.1049/tpe.1971.0034.

Fukuyama, M. (2018). "Society 5.0: Aiming for a New Human-Centered Society." Japan SPOTLIGHT 27 (Society 5.0): 47–50.

Ghobakhloo, M. (2019). "Determinants of Information and Digital Technology Implementation for Smart Manufacturing." International Journal of Production Research 1–22. doi:10.1080/00207543.2019.1630775.

Gloy, Y.S., et al., Book Textile Sci. Eng 5 (2015) p. 1.

Hayek, P. A., Salameh M. (2001). "Production Lot Sizing with the Reworking of Imperfect Quality Items Produced." Production Planning & Control 12 (6): 584–590. doi:10.1080/095372801750397707.

Hermann, M., Pentek T., Otto B. (2016). "Design Principles for Industrie 4.0 Scenarios." In 2016 49th Hawaii International Conference on System Sciences (HICSS), 3928–3937. IEEE. doi:10.1109/HICSS.2016.488.

Horvath, D., Szabo R. (2019). "Driving Forces and Barriers of Industry 4.0: Do Multinational and Small and Medium-Sized Companies Have Equal Opportunities?" Technological Forecasting and Social Change 146: 119–132. doi:10.1016/j.techfore.2019.05.021.

Kagermann, H., Wahlster W., Helbig J. (2013). Recommendations for Implementing the Strategic Initiative Industrie 4.0. Final Report of the Industrie 4.0 WG. doi:10.13140/RG.2.1.1205.8966.

Kamble, S., Angappa G., Neelkanth D. (2019). "Industry 4.0 and Lean Manufacturing Practices for Sustainable Organisational Performance in Indian Manufacturing Companies." International Journal of Production Research 1–19. doi:10.1080/002075 43.2019.1630772.

Keqiang, L. (2015). Made in China 2025. Beijing: State Council of China. http://english.gov.cn/2016special/madeinchina2025/.

Kerpen, D., et al. (2016). "Effects of Cyber-Physical Production Systems on Human Factors in a Weaving Mill: Implementation of Digital Working Environments Based on Augmented Reality." In 2016 IEEE International Conference on Industrial Technology (ICIT), 2094. Institute of Sociology (IfS) RWTH Aachen University, Aachen, Germany.

Kiel, D., Arnold C., Voigt K. (2017a). "The Influence of the Industrial Internet of Things on Business Models of Established Manufacturing Companies – A Business Level Perspective." Technovation 68: 4–19. doi:10.1016/j.technovation.2017.09.003.

Kiel, D., Meuller J., et al. (2017b). "Sustainable Industrial Value Creation: Benefits and Challenges of Industry 4.0." International Journal of Innovation Management 21 (08): 1740015–1740034. doi:10.1142/S1363919617400151.

Lasi, H., Fettke, P., Kemper, H.-G., Feld, T., and Hoffmann, M. (2014). "Industry 4.0", Business and Information systems Engineering 6 (4): 239–242.

Lee, J., et al. (2017). "On solving the inverse kinematics problem using neural networks." IEEE Internat. Conf. M2VIP 24, 1.

Li, C. Z., Hong J., et al. (2016). "SWOT Analysis and Internet of Things-Enabled Platform for Prefabrication Housing Production in Hong Kong." Habitat International 57: 74–87. doi:10.1016/j.habitatint.2016.07.002.

Li, Y., Ai J., Sun C. (2013). Online Fabric Defect Inspection Using Smart Visual Sensors, Sensors 13: 4659.

Liao, Y., Deschamps F., de Freitas Rocha Loures E., Ramos L. F. P. (2017). "Past, Present and Future of Industry 4.0 – A Systematic Literature Review and Research Agenda Proposal." International Journal of Production Research 55 (12): 3609–3629. doi:10.1080/00207543.2017.1308576.

Manglani, H. (2018). Web source, Available at: https://hmanglani.com/textile-automation.

Ministry of Enterprise and Innovation. (2016). Smart Industry: A Strategy for New Industrialisation for Sweden.

Muller, J. M., Buliga O., Voigt K. (2018). "Fortune Favors the Prepared: How SMEs Approach Business Model Innovations in Industry 4.0." Technological Forecasting and Social Change 132: 2–17. doi:10.1016/j.techfore.2017.12.019.

Ngjeqari, V. (2016). The Sustainable Vision of Industry 4.0. University of Vienna. https://50.
    unido.org/files/research-paper-competition/Research-Paper-Vojna-Ngjeqari.pdf.
OECD. (2017). The Next Production Revolution: Implications for Governments and Business.
    Paris: OECD Publishing.
OECD. (2018). Production Transformation Policy Review of Chile: Reaping the Benefits of
    New Frontiers. Paris: OECD Publishing. doi:10.1787/9789264288379-en.
Perez, J., et al. (2016). 'New Requirements for Higher Education.' Internat. Conf. Eur.
    Transnational Educ 126.
President's Council of Advisors on Science and Technology. (2014). Accelerating U.S.
    Advanced Manufacturing. Report to the President Accelerating U.S. Advanced
    Manufacturing. doi:10.1111/j.0033-0124.1964.033_g.x.
Reischauer, G. (2018). "Industry 4.0 as Policy-Driven Discourse to Institutionalize Innovation
    Systems in Manufacturing." Technological Forecasting and Social Change 132: 26–33.
    doi:10.1016/j.techfore.2018.02.012.
Saggiomo, M., et al. (2015). 'SozioTex-Sociotechnical systems in the Textile Industry:
    Interdisciplinary Competence Build-up in Human-machine Interaction Facing
    Demographic Change.' Melliand Int. 20: 49.
Saggiomo, M., et al. (2016). "Weaving Machine as Cyber-Physical Production System: Multi-
    Objective Self-Optimization of the Weaving Process." In 2016 IEEE International
    Conference on Industrial Technology (ICIT), 2084. IEEE.
Schneider, P. (2018). "Managerial Challenges of Industry 4.0: An Empirically Backed
    Research Agenda for a Nascent Field." Review of Managerial Science 12 (3): 803–848.
    doi:10.1007/s11846-018-0283-2.
Schumacher, A., Erol S., Sihn W. (2016). "A Maturity Model for Assessing Industry 4.0
    Readiness and Maturity of Manufacturing Enterprises." Procedia CIRP 52: 161–166.
    doi:10.1016/j.procir.2016.07.040.
Simonis, K., et al. (2016). IOP Conference Series: Materials Science and Engineering 141:
    012014.
Taleizadeh, A. A., Wee H., et al. (2010). "Multi-Product Production Quantity Model with
    Repair Failure and Partial Backordering." Computers & Industrial Engineering 59 (1):
    45–54. doi:10.1016/j.cie.2010.02.015.
Teknikforetagen. (2017). Made in Sweden 2030 – Strategic Agenda for Innovation in Production.
    http://www.teknikforetagen.se/globalassets/i-debatten/publikationer/produktion/made-
    in-sweden-2030-engelsk.pdf.
Viharos, F., et al. (2016). "Vision Based, Statistical Learning System for Fault Recognition in
    Industrial Assembly Environment." In Emerging Technologies and Factory Automation
    (ETFA), 2016 IEEE 21st International Conference on, 1–6. IEEE.
Wang, S., Wan J., et al. (2016). "Implementing Smart Factory of Industrie 4.0: An
    Outlook." International Journal of Distributed Sensor Networks 12 (1): 3159805.
    doi:10.1155/2016/3159805.
Winck, R., et al. (2009). The IEEE/ASME International Conference on Advanced Intelligent
    Mechatronics (AIM), 53.

2 Practical Issues in Robotics Internet of Things

Yogini Dilip Borole, C.G. Dethe,
Roshani Raut and Prasenjit Chatterjee

CONTENTS

2.1 OVERVIEW

Keen administrations are becoming progressively key with the development of the fourth phase of industrialisation, such as 4th Industrialised Rebellion, where new troublesome advances are evolving both modern and exploration fields. Through the primary three recent upsets, there has been an improvement of efficiency on account of the making of new mechanical, electrical and electronic innovations.

Over recent years, the requirement to improve human life quality has prompted figuring out increasingly more creation models of customised and computerised administrations [Lu, Y. 2017]. The principal consequence of the fourth modern unrest lies in the turn of events and quick spreading of the cyber-physical systems. Cyber-physical systems figure out how to create applications combining physical resources for computational abilities. Advances of cyber-physical systems spread a wide scope of utilisations, for example, electrical force networks, transportation frameworks, social insurance gadgets and gas conveyance. In cyber-physical systems, the cooperation with the physical frameworks integrates through

DOI: 10.1201/9781003122401-2

systems, running complex examination while separating information. The utilisation of system administration, web and sensing devices in cyber-physical systems prompts the meaning of Internet of Things (IoT) [Shih, C.S. et al. 2016]. IoT can be viewed as the foundation that makes cyber-physical systems conceivable, as IoT frameworks are based on correspondence conventions, on account of which physical resources figure out how to interface with one another, moving and trading data. While the IoT worldview isn't expected to break down information furthermore, data frameworks, cyber-physical systems are intended to misuse IoT correspondence design to run complex investigation through a unified examination center, where data can be extricated from crude information to send control orders to the physical resource.

The cooperation with IoT and cyber-physical systems establishes a solid framework for the advancement of another exploration region, which is known as Internet of Robotics Things (IoRT). This new idea has a few changes in various areas that also spread a few applications [Xu, H. et al. 2018], including frameworks that need to work in testing conditions. For example, IoRT frameworks can be embraced by assembling enterprises for self-ruling and distantly performing testing assignments, for example, gathering, bundling, welding and overseeing quality control. Additionally, IoRT frameworks are likewise spreading outside the business condition, in exhibition halls, sports and diversion [Kondratenko, Y.P. 2015]. Considering all things, their improvement is because of the need of making interconnected frameworks with regard to 4th Industrialised Rebellion [Zhou, K., Liu, T., Zhou, L. 2015; van Erp, T.; Seliger, G. 2016], combining computerised and physical universes. In this situation, IoRT speaks to the center of mechanical autonomy implanted IoT frameworks, where distributed computing and system administration can be executed to achieve explained undertakings, permitting robots to share, arrange, what's more, assemble various types of data among the two people and machines. A plan to apply autonomy, distributed computing and IoT coordination can be seen in Figure 2.1.

Numerous audits have been written over recent years [Vermesan, O. et al. 2017; Batth, R.S., Nayyar, A., Nagpal, A. 2018; Mazzara, M. et al. 2019; Yao, X. et al 2019; Afanasyev, I. et al 2020], yet none of them arranges the IoRT frameworks as indicated by shrewd spaces. This study is proposed to reveal insight into the IoRT applications related to various shrewd areas with regard to 4th Industrialised Rebellion, characterising the cutting edge advancements of IoRT and laying out how IoRT frameworks could assume a key job in our society. The primary objective is to plot the significant difficulties in each field, intending to comprehend in which zones the usage and investigation of IoRT applications must be additionally explored. As this original copy is for the most part centered around the mechanical and creation fields, IoRT applications in brilliant assembling and keen agribusiness are profoundly examined. Further spaces are then investigated, for example, human services, instruction and observation, to lay out how IoRT frameworks are spreading in numerous perspectives of regular day-to-day existence.

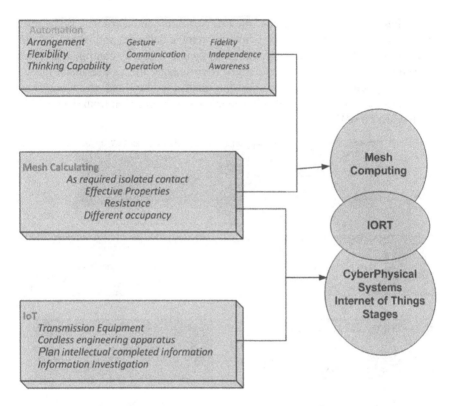

FIGURE 2.1 Features of Automation, Mesh Calculating, and IoT, which converge into the Web of mechanical equipment.

The primary commitments of this chapter include:

- An image of the connection among physical and virtual situations overseen by the cyber-physical system framework advances
- The meaning of the IoRT idea: its engineering and primary innovations
- The recognisable proof of different application spaces, plotting the most up to date best in-class writing in view of IoRT innovations
- Exposed matters and difficulties that merit examining later, demonstrating how IoRT frameworks could speak to a key job with regard to the fourth modern upheaval

The composition of this chapter is as follows: Section 2.2 outlines the principle highlights engaged with 4th Industrialised Rebellion, concentrating on advances dependent on the cyber-physical systems. Consequently, a layout of the IoRT is provided, analysing its parts and engineering. Section 2.3 characterises the most recent spaces and applications where automated frameworks are coordinated with IoT innovations. At last, Section 2.4 gives a conversation with respect to open issues and difficulties, while Section 2.5 reaches the determinations of the study.

The IoRT is a developed level of the IoT, permitting to incorporate such present-day advancements as distributed computing, remote detecting and impelling, information investigation, data support, direction, security, security control, appropriated observing and organising from the smart environment, just as decisional independence, recognition, control, multi-operator control, control and arranging, limitation and planning, route, swarm and human-robot collaboration from the robot side, which is shown in Figure 2.2.

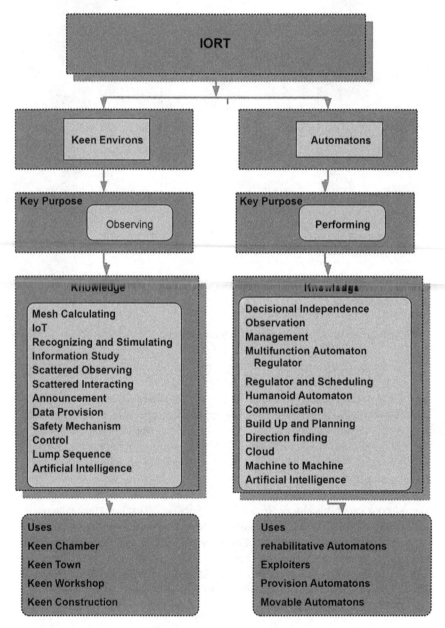

FIGURE 2.2 The IoRT building plan.

2.2 BRILLIANT TECHNOLOGIES IN 4TH INDUSTRIAL REBELLION

The fourth mechanical unrest, regularly known as 4th Industrialised Rebellion, can be portrayed as the following level of assembling, where advanced reconciliation and smart designing are utilised to change the way machines interact with one another and people. The development of 4th Industrialised Rebellion has prompted an improvement of the whole assembling industry [Wan, J. et al. 2016], bringing forth the purported smart spaces [Zheng, P. et al. 2018] what's more, more explicitly, smart factories. The principle capacity of smart spaces is to screen applications, forms, for example, power utilisation or condition of sensors and actuators, in a characterised controlling region. Specifically, the term smart factory alludes to another way to deal with mechanical assembling, where innovations in IoT are utilised to control the assembling procedure while gathering, breaking down, furthermore, trading data [Petitti, A. et al. 2016].

Carefully associated with the keen manufacturing plant is the idea of smart machine [Wollschlaeger, M., Sauter, T., Jasperneite, J. 2017]. These IoT-based machines incorporate regulators and add-on sensors, and utilise constant information from their own segments and different machines to accomplish mindfulness and self-examination [Leitão, P. et al. 2016]. Mindfulness empowers machines to make diagnostics in regard to conceivable breaking down segments, while self-correlation empowers the machine to design their settings appropriately, in view of their working history. For this, arrange models and correspondence conventions, both wired and remote, are utilised to facilitate sensors and actuators for more mind-boggling modern applications. Modern robots assume a key job in the Savvy Manufacturing condition, where tasks, for example, collecting, welding and splash painting are converged with IoT advances to ensure association among different gadgets and robots, guaranteeing dependable and proficient creation [Grieves, M., Vickers, J. 2017].

2.2.1 Digital-Physical Systems

Unique component identified with 4th Industrialised Rebellion is the purported digital-physical systems. With the nonstop improvement in various innovation areas, cyber-physical systems applications have become principal, as they figure out how to associate every physical gadget to the Internet, combining virtual, physical universes to accomplish keen items and creation [Tomiyama, T., Moyen, F. 2018]. The physical world, the Internet and correspondence systems are at the center of cyber-physical systems structure: (i) the physical world alludes to those physical items, procedures and conditions to be observed or controlled; (ii) the Internet speaks to those data frameworks, for example, administrations, applications and dynamic units; while (iii) correspondence systems allude to those segments that figure out how to interface the Internet with the physical world.

The utilisation of cyber-physical systems is in situations where machines sensors and actuators are connected to achieve the most noteworthy proficiency regarding creation, and the idea of digital-physical creation system establishes its framework with regard to the purported digital twin [Garcia, C.A., Montalvo-Lopez, W., Garcia, M.V. 2020], where physical modules are connected to virtual modules, expecting to make an association between physical components and the relating computer-ised rendition [Yu, M. et al. 2016]. With the approach of the fourth modern unrest, control and creation frameworks have increased a lot of significance in many savvy areas.

Especially, brilliant items and keen creation frameworks are carefully identified with the cyber-physical systems, and unique models have been intended to manage disappointments in discrete-occasion forms [Jiang, J.R. 2018]. Cyber-physical systems can be useful in improving the adaptability of IoRT-based creation frameworks in shrewd spaces, for example, producing horticulture, clinical medical procedure and senior consideration. In particu-lar, humanoid–automaton communication, where people and machines coop-erate, can be applied in these fields just referenced, especially in assembling. Here, the data given by the physical contact among people and robots can be utilised in frameworks, for example, gathering and welding. In such settings, characterising a Highly Adaptive Cybersecurity engineering is essential in the plan and usage of cyber-physical systems [Liu, C., Jiang, P. 2016], intending to influence the direction of a robot depending on focuses given by human specialists.

As cyber-physical system is the primary phase of advancement, it uncov-ers with key to characterise its structure and strategy. In modern settings, the engineering of cyber-physical systems is described by five unique levels [Liu, Y. et al. 2017], which can be seen in Figure 2.2. Such structure, known as five-layer design, figures out how to explain how to develop a cyber-physical system from the underlying information procurement to the last worth cre-ation. All things considered, a few cyber-physical system structures have been grown, each concentrating on various angle, which plan to all the more likely describe modern frameworks and keen processing plants with regard to 4th Industrialised Rebellion [Cai, Y. et al. 2019; Drozdov, D. et al. 2019]. In the accompanying, the levels of cyber-physical systems five-component engineer-ing are delineated.

Figure 2.3 is the primary phase of advancement of five-stage engineering of CPS executed in mechanical settings; it uncovers to be key to characterise its structure and strategy. In modern settings, the engineering of cyber-physical sys-tems is described by five unique levels [Liu, Y. et al. 2017], which can be seen in Figure 2.3.

Such structure, known as cyber-physical systems five-layer design, figures out how to explain how to develop a cyber-physical system from the underlying infor-mation procurement to the last worth creation. Considering all things, a few cyber-physical systems structures have been grown, each concentrating on various angles, which plan to all the more likely describe modern frameworks and keen processing plants with regard to 4th Industrialised Rebellion [Cai, Y. et al. 2019; Drozdov, D. et al. 2019]. In the accompanying, the levels of 5C engineering are delineated in Figure 2.3.

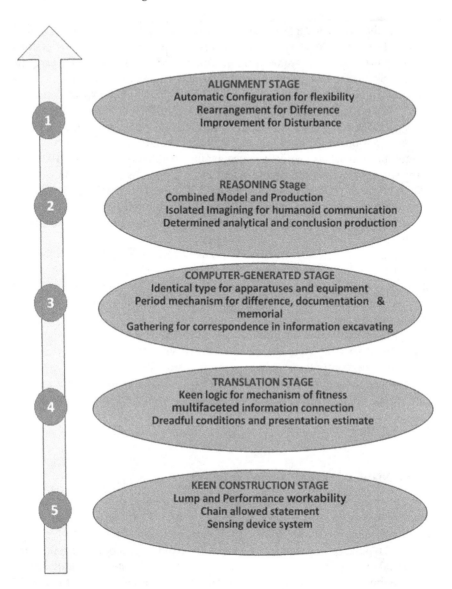

FIGURE 2.3 Five-stage engineering of CPS executed in mechanical settings. Lower levels gather information, which is investigated and dense at every upper level. With this arrangement, data taking care of to each more elevated level is more important than data coming into the level underneath.

The canny linking arrangements with the securing of precise and dependable information could be straightforwardly extricated from legitimate sensors or assembling frameworks. Such information is moved to the focal worker utilising explicit conventions, where the translation figures out how to separate significant data by methods for calculations and strategies that are created for the use viable. In the replicated glassy, an enormous measure of information is accumulated. Here, extra data is removed to give better knowledge on the status of each machine among the framework, planning to give machines self-correlation capacity. Such data is sent through appropriate designs in the cognizance level, where information about the machine status can be obtained, expecting to think about data and settle on the right choice to streamline the looking after procedure. At long last, the criticism from the Internet to the physical space happens in the conformation, going about as administrative control to make machines self-configurable and self-versatile. The developing need to utilise sensors and organised machines prompts the improvement of an enormous group of data, known as large information [Hosseini, M., Berlin, R.R., Sha, L.A. 2017]. Cyber-physical systems can be valuable to oversee both large information and also the interconnection among machines, making the industrial facilities reasonable for the 4th Industrialised Rebellion.

For the most part, cyber-physical systems comprise two principle highlights: (i) the limit of securing information progressively from the physical world, and a propelled network that figures out how to pick up data input from the Internet; (ii) computational force, which are coordinated into the Internet. Subsequently, cyber-physical systems can understand a powerful coordinated effort with physical frameworks. The last one gathers information by dispersed field gadgets in cyber-physical systems, which plans to ensure the continuous ability and precision of the gathered information. Cyber-physical systems have various uses, for example, modern control [Koken, B. 2015], disseminated vitality framework [Razafimandimby, C., Loscri, V., Vegni, A.M. 2020] and advanced clinical field [Ramson, S.J., Moni, D.J. 2017].

2.2.2 NETWORK OF MECHANICAL EQUIPMENT

As of late, the fourth modern transformation has prompted the improvement in the supposed Web of Robotics Things [Afanasyev, I. et al. 2020], which figures out how to achieve frameworks with dynamic self-rule, recognition and control. A layout of the IoRT parts can be found in Figure 2.4. The most exceptional idea as far as mechanical technology lies in IoRT, where cyber-physical systems are utilised to lay a solid establishment in improving the Internet of Things itself [Yao, X. et al. 2019]. In IoRT frameworks, present-day mechanical advancements have been converged with distributed computing [Nathalie, M., David, S.R. 2013] and system administration, coordinating cyber-physical systems and Internet of Things conventions. As a result of such combination, shrewd gadgets become ready to screen occasions, consolidate sensor information from an assortment of sources, and utilise nearby and circulated insight to decide the best strategy [Li, T.S. et al. 2017]. This new methodology highlights various innovations to perform confounded undertakings and work in a heterogeneous situation [Batth, R.S., Nayyar, A., Nagpal, A. 2018]. Besides, wireless sensor networks are turning into a matter of extensive significance in the most recent years. Wireless sensor networks are broadly researched by Cecil, J. et al. [2019],

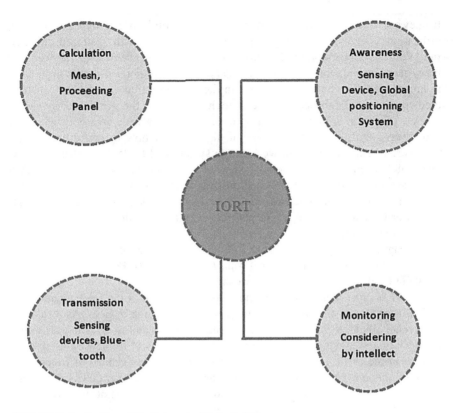

FIGURE 2.4 Outline of the Internet of Robotic Things components.

taking into account what number of new difficulties they suggest, and a number of new approaches have been extended to address them. In particular, the need of conveying sensor hubs as indicated by specific calculations and circumstances prompts the chance of incorporating such sensing devices with machine systems [Ke, Q., Xiaogang, L. 2016]. Such automatons live together with sensing devices, participating to upgrade the capability of the wireless sensor network. Therefore, the usage of machine systems with remote sensing devices has added to the improvement of new IoRT-based innovations and applications that can address various assignments, from sending to correspondences. Motorised frameworks have acquired impressive changes to different parts of human life. Both in the machine-driven and the scholarly universes, automatons have been utilised in playing out a wide range of complex and testing assignments, for example, bundling [Rosenfeld, A. et al. 2017], collecting [Gaggero, M. et al. 2019] and welding [Pai, V., Shenoy, U.K.K. 2019]. In this specific circumstance, IoT and powered skill combine for the improvement of new heterogeneous automated frameworks, with the point of improving the self-governing conduct of the robots. Also, the joining among mechanical autonomy and organising is of basic significance in the advancement of the IoRT frameworks. Organised robotics combines robot framework models (both equipment and programming) and applications that utilisation systems, for example, the web technology with scattered calculating [Vos, M., Cranefield, J. 2017]. As of late, the advancement of the web and many more applications in autonomy has prompted organised robots to be used in numerous fields, becoming central in the reconciliation of machines, distributed computing, databases, and even people throughout the world, as they can be in view of a local area network or appropriated over a wide area network. IoRT frameworks endeavor such innovation, figuring out how to coordinate robots with shrewd sensors that can work in the system and trade data. Officially, the term IoRT alludes to a biological system of keen gadgets ready to screen occasions, assemble and dissect information from various sources, trade data, utilising nearby and disseminated knowledge to decide an ideal grouping of activities and afterward act to change the physical condition while genuinely traveling through that condition.

The significant engineering of the IoRT frameworks, as portrayed in Figure 2.4, comprises three fundamental layers [Wan, J. et al. 2016]: (i) PHY level (ii) web and regulator levels and (iii) facility and use layer.

The PHY layer identifies to the most minimal degree of the IoRT engineering, which incorporates machines, sensing devices and moving devices. As a rule, the word robot alludes to vehicle, drones, automated vessels and so on. The words sensing devices and moving devices again allude to any sort of framework used to see and act in the earth, individually, considering a wide arrangement of frameworks going from home machines to modern sensing devices. Machines collaborate to create multi-machine frameworks, which incorporate various operators working in a similar situation. Such innovation can create and redesign anway applications with respect to automated frameworks, distantly oversee dispersed exercises and increment flaw resistance to accomplish an improvement in the general framework exhibitions [Belbachir, A., Benabid, S. 2016]. Moreover, sensing and moving devices can be incorporated into machine uses, expecting to upgrade, screen and control different procedures, for example, route, alignment and tuning [Mosavi, A. Varkonyi-Koczy, A.R. 2017].

Considering all things, a total and proficient coordination of sensing elements and moving devices with machine applications occurs at the system and controller level, where various parts can depend on specific conventions to impart and control forms. The second layer of the IoRT design incorporates regulators, switches, workers and different correspondence and control conventions. Specifically, correspondence procedures, for example, wireless fidelity, ZigBee protocols [Kehoe, B. et al. 2015; Alcácer, V., Cruz-Machado, V. 2019], Bluetooth small vitality, smart tags [Yin, S. et al. 2015; Alcácer, V., Cruz-Machado, V. 2019], nearby arena transmission and wireless sensor network based [Pedersen, M.R. et al. 2016; Alcácer, V., Cruz-Machado, V. 2019] are utilised to accomplish smooth data communications among automated frameworks of both short and significant distances as shown in Figure 2.5.

At long last, The service and application layers make extensive use of IoRT engineering, which emphasizes the usage of projects to control, process and

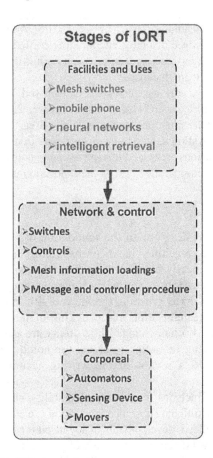

FIGURE 2.5   Chart of IoRT scheme design.

break down both natural boundaries and specialists (automatons, sensing devices and movers) in shrewd conditions. Moreover, this layer incorporates AI with neural network calculations [Coronado, P.D.U. et al. 2017], which can ensure level combination among mechanical frameworks and IoT applications, to achieve upgraded answers for complex issues in bodily situations.

2.3 BRILLIANT FIELDS AND CLAIMS IN THE INTERNET OF ROBOTIC THINGS SCHEMES

With the fourth mechanical insurgency, where apply for autonomy, cyber-physical systems and cloud advancements are combined, various areas are profiting by quick advancement [Kumar, A. 2018]. In this unique circumstance, the IoRT frameworks provide a few focal points over conventional automated applications, for example, offloading calculation concentrated errands on the cloud, getting a lot of infor- mation and sharing data with different robots, planning to take in new abilities and information from one another [Ali, M.H. et al. 2018]. Additionally, IoRT applications can likewise be utilised distantly, encouraging crafted by the two analysts and modern administrators, and making it more open, permitting collaboration between people even from significant distances.

In the accompanying, various areas are characterised, in which IoRT arrangements have been drawn closer and also executed. Figures 2.1–2.5 in the accompanying subsections lay out the situations canvassed in this area, featuring the assortments of the most recent innovation tended to in various conditions. However, the accumulated writing covers various areas, and it is important to note that the methodological approach is frequently comparative, just as the innovation utilised.

2.3.1 ASSEMBLING

Machine-build creation develops to the foundation of shrewd assembling, and the idea of mechanical machines has been happening a nonstop change in the most recent years, primarily because of the installing of IoT advances [Chang, W. et al. 2019]. IoRT develops to the primary empowering agent of such change, through which fabricating is grasping the ideas of 4th Industrialised Rebellion, implanting sensors, mechanisation and checking of items and forms [Liu, Q. et al. 2020]. Industry 4.0 has changed how items are made, adjusting to such mechanical development, intending to create top-notch merchandise and enterprises. Table 2.1 shows a review of the writing accumulated identified with shrewd assembling.

Shrewd assembling includes framework adaptability, checking and variation to change. Particularly, additional substance fabrication is a straightforward method of assembly strategies. Indeed, developments in advanced innovations that are happening during the Industry 4.0 need to stay aware of progressions in assembling procedures and materials [Lee, Y.K., Goh, Y.H., Tew, Y. 2018]. In this situation, with the point of encouraging brilliant assembling by sensing device frameworks, adaptable gadgets of added substance fabricating and their unwavering quality during forms are of basic thought.

IoRT may remember a few applications for the assembling area, for example, commercial joining and splash canvas [Qing, G., Zheng, Z., Yue, X. 2017]. In particular, commercial joining alludes to apply a welding instrument to a specific article, for example, a vehicle body, at determined discrete areas [Chen, X., Qin, S. 2017], while shower painting includes covering a surface with an indeed, even layer of paint, predetermining the direction along which the machine will move [Xin, L. et al. 2018]. The two applications may require ongoing adjustment of the way to oblige little deviations from the normal way, intending to accomplish satisfactory creation and stay away from issues, for example, incorrectness and impacts with both people and machines. As the capacity to perform self-limitation and exploring is of critical significance for a few modern machines, way arranging is as of now an examination challenge in the assembling space [Grau, A. et al. 2017; Mrudul, K., Mandava, R.K., Vundavilli, P.R. 2018; Yao, B. et al. 2018].

TABLE 2.1

Outline of the Writing Considered for the Industrial Area

Summary of the Workings	Automaton Steering with Track Scheduling	Information Assembly	Machine Vision	Mesh Figuring	Multifarious Machines	Human Robot Interaction
Execution of a dream framework on a machine controller to extend the capability of the incorporated photographic–machine framework in mechanical uses.		✓	✓			
Programmed way arranging of a six-hub robot controller for savvy producing, utilising a cloud stage that screens the framework through transmission control protocol/Internet protocol convention for organised far-off controlling and recreation.	✓	✓	✓			
Coordination of machines in *Certified Professional in Patient Safety* to oversee diverse weight products, joining unmanned ground vehicle with machine controller and air-move frameworks to manufactured keen processing plant and shrewd fabricating.	✓			✓		
Digital-physical self-governing versatile machine was able to do performing human–robotic interface by permitting clients to oversee orders utilising a cloud stage. The robot moves following the arranged course map, as per the impediment evasion framework, until it arrives at the goal and informs the cloud stage.	✓			✓		✓
Precise improvement structure, utilised for humanoid–machine cooperative dismantling in practical assembling. An expandable observation stage for modern machines framework and human body is characterised by methods for a honey bees calculation-based succession arranging technique for a handheld remote controlled device task.		✓		✓		✓

(Continued)

TABLE 2.1 (Continued)
Outline of the Writing Considered for the Industrial Area

Summary of the Workings	Automation Steering with Track Scheduling	Information Assembly	Machine Vision	Mesh Figuring	Multifarious Machines	Human Robot Interaction
Way arranging calculation, utilising quick walking strategy for a human automaton to move in a stationary situation, planning to let it move in both known and obscure situations.	✓	✓				
Portable robot way arranging, consolidating unconventional search, and Flicker calculations to achieve the ideal way.	✓	✓				
Model of a framework for pressing grouped treats, building up a structure to associate shoppers, brilliant production lines and different frameworks through mesh and strategic systems.				✓	✓	
Blunder design change depending on repeat nearest opinion calculation for entity present estimation of a machine controller, utilising point mesh information assembled from different sound system vision frameworks.	✓	✓	✓			
Touchless outer power location delivered by human administrators in physical human–robotic interface, meaning to acquire a dynamic model of a mechanical machine controller in both dynamic and semistationary manner.					✓	✓

Moreover, modern controllers that gave visual direction can additionally improve the way arranging frameworks, as a dream-based methodology provides more attention to identify and stay away from hindrances [Madhira, K. et al. 2017; Liu, Q. et al. 2019]. The fundamental concern with respect to way of arranging is the negative effect on the computational speed of the way of calculation that can be brought about by enormous information stockpiling and complex registering forms. Consequently, it is important to improve the effectiveness of the way of arranging calculation just as incorporate a distributed computing approach, intending to oversee equipment asset limitations of modern machines [Wan, J. et al. 2016].

With distributed computing, task executions are built up conjuring cloud advancements, guaranteeing coordination and interoperability [Zimmerman, M., Marvel, J.A. 2017]. In assembling, the gathering procedure has additionally changed in the fourth industrial replication period, advancing into the mechanical sequential construction system, where robots figure out how to perform errands recently performed by human work [Halme, R.J. et al. 2018]. The gathering procedure compares to mate unique parts with the point of building up another subsegment, or a completed item. IoRT advances can improve such procedure, exploiting implanted CPSs [Saraswathi, M., Murali, G.B., Deepak, B. 2018; Wang, S. et al. 2018], planning to utilise numerous machines to permit parts to be re-situated or controlled, executing more perplexing get-together techniques. In addition, the presentation and improvement of cloud mechanical autonomy have totally changed and improved the adaptability and extensibility of errand planning for assembling [Garciaa, M.A.R. et al. 2018]. By interfacing with the cloud, modern automatons become ready to download information with respect to the assembling engineering and furthermore, different automatons remembered for the framework, figuring out how to defeat a few confinements about data and learning [Duckett, T. et al. 2018]. Along these lines, the computational intensity of the cloud workers redresses for the constrained computational intensity of the mechanical framework itself. For instance, Romero-Hdz, J. et al. [2016] characterise a novel mesh-assisted and self-sorted out assembling framework to incorporate the assembling applications in the cloud worldview. All the more explicitly, the work is centered around a customised sweets pressing application, interconnecting physical shop floor substances, for example, automatons, transports and items, furthermore, customer terminals. During the cooperation procedure, a lot of information are gathered into the cloud terminal, and furthermore, application programming is conveyed to process such information.

Another essential part of savvy producing includes communitarian machines, which are reforming the mechanical autonomy showcase. Communitarian automatons are protected and instinctive to utilise, figuring out how to help human administrators, adjusting to vulnerability [Wolfert, S. et al. 2017]. For this reason, the area of the human, alongside the area of her/his appendages, must be known progressively, while the robot itself moves. Inertial sensor-based strategies and vision-based sensors can be helpful to recognise human nearness inside community-oriented workspaces [Saravanan, D., Archana, G., Parthiban, R. 2018]. In writing, a few regions of exploration manage humanoid–machine cooperation, intending to create robots appropriate in a specific workplace [Kalpana, M.A.T.M., Tyagi, M.A. 2017; Chen, F., Selvaggio, M., Caldwell, D.G. 2019], exploiting sensing devices for assembling circumstance mindfulness. The reconciliation of such IoRT frameworks in savvy producing is proposed to use the qualities of the two robots and human administrators, intending to make up for the restrictions of each other [Prasad, D., Singla, K., Baggan, V. 2019].

2.3.2 HORTICULTURE

Trend setting innovation in 4th Industrial Rebellion has obtained a significant development farming. Work-escalated errands, for example, the gather of new leafy foods, are encountering a generous change because of the coordination of IoT-based frameworks in the agribusiness space [Li, X. et al. 2019]. Despite the fact that most ranchers are as yet dependent on customary strategies, as of late, there has been critical intrigue in creating shrewd farming frameworks, concentrating on issues, for example, trimming yield, weed control and information gathering from fields [Xu, X., Li, X., Zhang, R. 2019]. In this unique situation, IoRT frameworks can be structured and actualised to enable homesteads with advancements that coordinate administrations, items and information so as to increment efficiency, quality and benefit, exploiting the coordinated effort among people and brilliant frameworks as shown in Table 2.2 [Zaman, S. et al. 2019].

The specific weakness of farming to environmental change prompts the essential savvy horticulture ideas: weather-based intelligent cultivation and supportable strengthening [McConnell, M.D., Burger, L.W. 2017]. The fundamental objective of weather-based intelligent cultivation is to build finance and food security while diminishing greenhouse gas discharges [García-Tejero, I.F. et al. 2020]. As environmental change has a huge and by and large negative effect on horticulture, ranchers and homestead specialists' co-ops must react viably in the long haul for such issues, just as having the option to deal with the hazard related to expanded atmosphere changeability. In this unique situation, on-ranch water stockpiling and water system, just as overhauls of ranch ventures themselves, are of basic thought, for what it's worth central structure agrarian frameworks and ready to improve their own dependability through dust, liquid, and plant compost the board [Tyagi, N.K., Joshi, P.K. 2019]. For this reason, *Cloud Security Alliance* frameworks dependent on the IoRT engineering permit editing assortments that are more open minded toward heat, dry seasons, floods and saltiness, and are expected to go way past the basic objective of escalating horticulture [Uddin, M.A. et al. 2017]. *Cloud Security Alliance* and significant integration are carefully associated with one another, as the two address environmental change centering on expansion and misusing complementarities

TABLE 2.2
Sums Up the Writing Considered for This Segment

Summary of the Workings	Automaton Steering with Track Scheduling	Information Assembly	Machine Vision	Mesh Figuring	Multifarious Machines	Human Robot Interaction
A few unmanned aerial vehicles are utilised to gather information by observing and planning the field to change rate compost, splashing, and so forth, to lessen crop maladies.	✓	✓	✓			
Portable automaton outfitted with a few sensing devices helpful in farming (dampness device, temperature device, pollution device, harm of gather device), and constrained by voice acknowledgment, utilising a savvy associated with the system.	✓	✓				✓
District observing of plants in a savvy nursery, utilising a cloud-helped procedure of versatile robots to build the checking area measure and diminish time utilisation.	✓	✓		✓		
Distantly configurable yield picture procurement machine framework, in view of distributed computing and wireless sensor network, used to improve the adaptability and transformation of the portable machine.		✓	✓	✓		
Continuous picture handling calculation, utilising a visual odometer framework on an unmanned ground vehicle, in view of the cross-relationship approach. Low-goal pictures are utilised to accomplish high exactness moving estimation with short processing time.	✓	✓	✓			
Participation among mixed farming field automatons with an administrative regulator, utilising an innovative method dependent on distinct-occasion framework and the supervisory control theory hypothesis, which is compelling in controlling complex unique frameworks comprising mixed multifunction humanoid for shrewd farming.	✓	✓			✓	

(Continued)

TABLE 2.2 (Continued)
Sums Up the Writing Considered for This Segment

Summary of the Workings	Automaton Steering with Track Scheduling	Information Assembly	Machine Vision	Mesh Figuring	Multifarious Machines	Human Robot Interaction
Keen agriculture framework dependent on inserted gadgets, IoT and wireless sensor network for agriculture-ranch stock and animals ranches.	✓	✓				
Unmanned ground vehicle utilised for searching for the best appropriate conveying position for a Wireless Sensor Network framework, intending to dissect the field also, assemble data about the landscape condition.	✓	✓				
Computerised framework created to switch both atmosphere and water system in a nursery by observing heat, dust dampness, mugginess, and potential of hydrogen, utilising a mesh-associated versatile machine. Such machine can likewise find undesirable plants utilising picture preparing.	✓	✓	✓	✓		
Organisation of a gathering of unmanned ground vehicle utilising a disseminated calculation, meaning to assemble information from significant regions of the field, chose utilising the polygonal partitions apportioning.		✓	✓		✓	

between crops. Since ranchers won't embrace research for environmental change transformation that may not yield improved profits for interests in the present moment, significant integration occurs that addresses asset shortage and natural difficulties, expanding reaping and arable harvests, intending to locate the best way to deal with keeping up the compromise among yields and natural needs in various conditions. While in certain regions, yields are perfect with ecological enhancements, in different situations yields decrease or land reallocation can be fundamental so as to guarantee supportability and assurance advantages, for example, natural life preservation [Serebrenny, V., Shereuzhev, M., Metasov, I. 2018], fuel stockpiling [Mekala, M.S., Viswanathan, P. 2017] and overflow security [Reina, G. et al. 2016]. To this point, information examination accumulated by methods for IoT frameworks [Milella, A. et al. 2019] and portable robots, joined with distributed computing, can give viable data with respect to the degree of water assets, stickiness and synthetic concoctions [Khatri-Chhetri, A. et al. 2017]. In addition, picture acknowledgment and preparing methods, for example, crop picture procurement, can additionally disclose such data. In the IoRT frameworks, crop picture procurement design, in view of distributed computing and remote system, can build adaptability and variation of portable vehicles [Mat, I. et al. 2018; Romeo, L. et al. 2020], conquering the conventional picture, securing frameworks which primarily depend just on a fixed camera or the portable robot itself.

In farming, mechanical frameworks can be conclusively useful in accomplishing both high caliber and item amounts, as human abilities and agrarian hardware are profoundly restricted concerning machines' aptitude. The procedure of mechanical reconciliation in agrarian situations has prompted the robotisation of those apparatus that assume a key job in horticulture, for example, water system and manure frameworks, reapers and farm vehicles [Petkovics, I. et al. 2017]. The reconciliation of IoRT frameworks in agrarian apparatus requires another methodology to deal with the control signals from the control framework toward the movers. Such frameworks ought to ensure an improvement in the monetary feasibility, diminishing ecological effect and expanding food supportability [Li, X. et al. 2019].

IoRT-created savvy horticulture frameworks are intended to perform different farming exercises, such as dampness detecting, water system, crop observing and guard against bothers and unsafe creatures [Valecce, G. et al. 2019]. For this reason, versatile automatons with coordinated radio direction finder, for example, territorial and crushed vehicles [Dharmasena, T. et al. 2019; Mahbub, M.A. 2020], are utilised to gather information from the field [Jawad, H.M. et al. 2017; Basnet, B.; Bang, J. 2018]. Automated aerial vehicles are regularly utilised alongside a lot of IoT gadgets, to accumulate data from such sensing devices, overflying the region [Scilimati, V. et al. 2017].

Accuracy farming is done to utilise an IoRT to deal with herbicide, manure, or water systems, expecting to oversee various degrees of, for example, varieties of harvest sizes, light and climate conditions [Faryadi, S., Davoodi, M., Mohammadpour Velni, J. 2019; Ju, C., Son, H.I. 2020].

In the primary objective of this homestead, the board approach is to productively lessen ranch assets, constraining the cost of horticultural creation while augmenting the yield. To arrive at this objective, a system of smart sensor can be utilised to screen and measure any adjustment in plants, utilising a system of astute sensing devices [Campbell, B.M. et al. 2014; Blaser, W.J. et al. 2018]. In this specific situation, the wireless sensor network framework, made up of a set number of hubs coordinating radio recurrence phones, minor scales controllers, sensing devices and power foundations [Adamo, F. et al. 2017], can provide further help in collecting, controlling and checking of information [Krishna, K.L. et al. 2017; Micoli, G. et al. 2019]. IoT gadgets are conveyed in a particular region, intending to screen ecological boundaries [Roopaei, M., Rad, P., Choo, K.K.R. 2017], utilising a remote association with sending information naturally through multi-jump correspondence. Then again, because of the restricted independence of the gadgets, it is central to send the hub sensing devices mixed, intending to have the most extensive inclusion with least vitality use [Cao, H. et al. 2018].

WSN applications are valuable in brilliant agribusiness, as their primary highlights incorporate self-arranging, conclusion and self-sorting out [Grigorescu, S.D. et al. 2019]. Such properties offer high spatial and transient goal to screen crops through the device hubs sent over the rural fields [Amin, R. et al. 2018]. In the study of Campbell, B.M. et al. [2014], a sensing device arrange is consequently sent by a portable automaton, which is utilised to search for the best appropriate position of the conveying hub sensors, expecting to assemble; however, much information could reasonably be expected from the encompassing condition. A correlation with a manual organisation has been directed, demonstrating that the automated supported methodology prompts better, figuring out how to expand correspondence proficiency, expecting to screen ecological boundaries in wide open-air conditions. The sending of an IoT arrange ment in keen horticulture, besides different advantages, can confine altogether the support costs.

Such agrarian frameworks, can actually provide information by methods for sensing devices for estimating heat, dampness, synthetic substances and so on [Zhou, H. et al. 2019]. With such information, aquatic, dung and insecticides can be self-governing conveyed through a mechanical framework, in more exact amounts and positions, and with better time booking to expand yields and reduce prices.

2.3.3 SUPPLEMENTARY AREAS: WELLNESS PROGRAM, TRAINING AND OBSERVATION

IoRT frameworks are considered to have a tremendous incentive in medicinal services, instruction and reconnaissance spaces. Table 2.3 involves an outline of the writing considered for this segment. In particular, IoRT can offer well-being, cultural and financial advantages in a few applications, specifically for those with specific needs, for example, mental handicap, stroke patients, victims and amputees [Chen, M. et al. 2018; Nayyar, A. et al. 2018]. The reconciliation of robots with sensors and IoT gadgets provides a few points of interest by giving constant

TABLE 2.3

Summary of the Works Measured for the Wellness Program, Training and Observation Area

Summary of the Workings	Automaton Steering with Track Scheduling	Information Assembly	Machine Vision	Mesh Figuring	Multifarious Machines	Human Robot Interaction
Mesh- and IoT-assisted indoor automaton for conveyance medication, in view of the multicenter installed framework, radio-frequency identification and standards from IEEE correspondence convention, and mesh stages.	✓			✓		
Engineering and plan of a wearable full of feeling robot furnished with psychological registering, named Fitbit. Such automaton can perform multi-modular information observation, intending to perceive the feelings of the sufferer.		✓		✓		✓
Intended to help home-driven human services framework, acknowledged by coordinating human-movement catch subsystem with machine-controller system. The automaton utilised is a double arm agreeable machine, which mirrors human movement caught by a lot of serviceable inertial movement catch gadgets to finish task.	✓	✓				✓
Acknowledgment of a well-being appraisal stand by creating a mechanical stage that guarantees its usefulness inside the Keen Urban data and correspondence systems, and, furthermore, can give explicit capacities by creating applications as indicated by the requirements of the sufferer.	✓					✓
IoT is dependent automaton framework prepared with both long-range and short-extend correspondence frameworks. The robot is productive in checking ongoing conditions for brilliant observation.	✓	✓				

(Continued)

TABLE 2.3 (Continued)
Summary of the Works Measured for the Wellness Program, Training and Observation Area

Summary of the Workings	Automaton Steering with Track Scheduling	Information Assembly	Machine Vision	Mesh Figuring	Multifarious Machines	Human Robot Interaction
Advancement of a portable reconnaissance camera observing framework, utilising a line supporter to give a portable development, planning to beat the constrained inclusion issue looked by customary reconnaissance cameras.	✓		✓			
Multifarious machines framework is dependent on group knowledge for reconnaissance and salvage missions, with continuous information transferring on cloud utilising IoT, abusing remote intercommunication between different operators, procedure and subterranean insect province advancement calculation, with the goal that they can achieve errands simultaneously.	✓	✓	✓		✓	
Observation robot utilised for climbing both even and vertical surfaces, while naturally controlling surface changes, investigating space and communicating live video through remote channel to the distant workplace.	✓	✓				
Terrestrial coalmine site and poisonous gas detecting utilising a multifarious field reconnaissance robot. Esp8266 with processor is utilised to line the controller and do machine, which can get on any landscapes, gathering data. All mechanical-sensing devices information are sent to mesh workers.		✓		✓		
Self-ruling networked automatons for reconnaissance, in which a wireless sensor network is actualised, where every sensing devices hub contains burn, ultraviolet intensity, scent and movement indicator sensors and radio-frequency handsets for system administration and correspondence.	✓	✓			✓	

well-being data and diagnosing and understanding conditions, planning to diminish the hazard of human missteps, for example, diagnosing incorrect medication portions, and techniques [Saad, W.H.M. et al. 2018]. Likewise, the IoRT advancements can acquire numerous focal points to different applications, for example, patients, staff and rescue vehicle, programmed information assembling and detecting [Singh, R. et al. 2018].

Concentrating on the instruction space, automatons require utilising proper and versatile practices to achieve and maintain sufficient social collaborations with individuals, to obtain help, for example, with schoolwork and instructing [RajKumar, K. et al. 2019]. Estimating the electrodermal movement can be utilised for the human–robotic interface, as it relates to an adjustment in the coating communication in light of scenes of consideration, nervousness and fervor [Raju, H.S., Shenoy, S. 2016]. Moreover, exploratory data analysis reactions estimated in youngsters can vary impressively from the normal reactions estimated in grown-ups [Meghana, S. et al. 2017], as youngsters may not react to certain driving forces as grown-ups do. In such circumstances, the IoRT advances utilised for youngster instructive concerns need to gather, process and investigate information, so as to accomplish programmed forecast of kids conduct and state. The monitoring of locations and individuals with the goal of ensuring human safety and safeguarding specific situations is a well-known practise. Here, IoRT frameworks assume a key job, as they can give brilliant advances for high observation in situations, for example, touchy regions, medical clinics, military fringes, open places and homes [Bonaci, T. et al. 2015; Ashok Kumar, M., Thirumurugan, T. 2018]. For the most part, close-circuit television and cameras are utilised for management of both indoor and open-air environmental factors [Shah, M.S., Borole, P.B. 2016]. In any case, such innovation involves a few issues and limitations, for the most part brought about by possible altering and the nearness of vulnerable sides.

Such disadvantages can be mostly illuminated by expanding the quantity of cameras in the framework [Grieves, M., Vickers, J. 2017], covering more alcoves, yet expanding the expense and the unpredictability of the framework itself. The coordination and advancement of the IoRT applications in reconnaissance seem, by all accounts, to be the most reasonable decision in observation settings, as the IoRT frameworks can be customised and executed to be quick and work productively, covering bigger regions so as to make secure a specific space [Chukwuemeka, C., Habib, M. 2018]. Likewise, cloud advancements incorporated with mechanical technology, which guarantee continuous discovery, are generally valuable to distantly screen conditions [Saponara, S., Pilato, L., Fanucci, L. 2016], for example, homes, businesses [Fierro, R., Chaimowicz, L., Kumar, V. 2018], retail and discount stores, as well as to distinguish human nearness in various situations [Talebpour, Z., Martinoli, A. 2018]. Sensor readings, for example, global positioning system, attractive field, nature of air and ecological qualities can likewise be useful in the management of both indoor and open-air situations [Fenning, R.M. et al. 2017; Khakimov, A. et al. 2017], as the automaton can communicate ongoing information during an observation strategic.

2.3.3.1   IoRT Architecture: Existing Robots

Generally, robots are divided into two categories: service robotics and field robotic [Pedersen, M.R. et al. 2016]. Robotic service stands for humanoid and domestic robots emanate to human's support functions, for example, homework, work, personal assistants for transportation, room cleaning and delivery.

On the other hand, field robotic identifies robot so operating in restricted and unregulated areas, especially outdoors. Field robots may be classy-being organised into three categories, aerial, marine and low traffic lights. These are the types of robot's open wide working and environmental conditions. Grieco et al. [Coronado, P.D.U. et al. 2017] list the different types of robots as shown in Table 2.4. Table 2.4 introduces the most important and relevant products for sale that belong to both service robotics and the camp robotic categories that can be used in combination with IoRT. Table is sorted by type, model, description and different uses and assigned "X" to appropriate application of robotic behavior. Details can be found in the customised datasheets directions that are given to each side. More equipment such as sensors, RFID, position constraints, weather, and cameras can be accepted by the robots in order to acquire additional information from nature, such as location, the presence of obstacles, people, nature, and objects, as illustrated in Table 2.5 [Coronado, P.D.U. et al. 2017].

2.4   PROBLEMS AND CHALLENGES

Safety is a mind-boggling issue in the region identified with both the IoT and automaton association. Principle network safety issues in mechanical autonomy can be caused because of the accompanying reasons underneath:

- Insecure correspondence among clients and machines leads to digital assaults. Programmers can without much of a stretch hack into unreliable correspondence connect in a matter of moments.
- Authentication issues. Disappointment in guarding against unapproved access can without much of a stretch permit programmer to enter the machine frameworks and utilise their capacities from distant areas without utilising any substantial username and secret word.
- Lack of appropriate encryption next to vendors that can open touchy information to likely programmers.
- Most of the robot highlights are programmable and reasonable. In the event that the default robot design is feeble to hacking, interlopers can undoubtedly gain admittance to the programmable highlights and change them.

The issue of network safety that can be real in mechanical technology is laid out by Liu, Y. et al. [2017], which talks about existing bugs and weaknesses that concede robots to be hacked distantly and applications that require security and protection to be actualised in the field of mechanical technology. Network safety is additionally pertinent for machines and mechanisation frameworks that depend on information and programming code from the system to continue their

TABLE 2.4

Existing Robots Are Considered for the Construction of IoRT

Category	Prototypical	Skills Explanation	Strength	Manufacturing Structure	Armed	Release Structure
Humanoids and home robots	Adept Mobile Robots People bots	Support for interactive robots and other activities related to telepresence, robotic vision, tourism, monitoring and control and education	X	X		
	Fraunhofer IPA-O-bot maintenance	Help for people in their daily lives	X	X		
	Willow Garage PR2	Work and home support services (including assistance for people with disabilities and the elderly)				
	IPAL Robotic REEM	Support for human services in many types of housing (i.e. hotels, museums, industry, shopping malls, airports, hospitals, care centers)	X	X		
	IPAL Robotic REEM	Support for human services in many types of housing (i.e. hotels, museums, industry, shopping malls, airports, hospitals, care centers)	X	X		
Robosoft Robulab's family	Turtle bot	Domestic infrastructure management, environmental awareness, liaison with medical and community facilities, key signal management, emergency call making, lifting and transporting people	X	X		
	Neobotix family vertical	An automated system for transforming industrial areas		X		
	Robotnik Automation Guardian	Ordinary robots. They can travel extensively and carry heavy loads. each device can be customised with sensors, grippers and GPS interface			X	X
Flying planes	AseTec Quadrotor	Environmental management and monitoring			X	X
Marine Robots	Explicit Robotics Kingfisher	Marine management and transport			X	X

TABLE 2.5

Available Robotic Machines Designed for IoRT Construction

Category	Prototypical	Explanation
Grade 2D Laser Finder	Hokuyo Scanner Range Finder Patients Laser LMS-2xx	Environmental recognition, determining human body size and position, identifying invaders and obstacles
3D sensors	Simulation of the Swiss Mesa Ranger Microsoft Kinect 3D Laser Prediction	Real-time production of high-quality 3D images during airline measurement terms
Cameras	3D Laser	Prediction and avoidance
RFID	RFID UHF RFID Reader	Identifying items and people
Position balancing	Applanix POS-LV Mode/GPS	Position gauge pose, or under extreme GPS conditions
LIDAR	RPLIDAR A2 360° Laser Scanner	Laser scanner for target distance by laser light

usefulness. This issue too influences huge information handling and distributed computing because of access to libraries, datasets, maps and so forth and cloud tasks that can likewise be related with access to resemble framework processing utilising on-request factual investigation, and consequently they ought to be remembered for the security umbrella [Liu, C., Jiang, P. 2016].

IoT-constructed applications for mechanical autonomy require explaining a few issues, creating procedures and picking structural arrangements [Drozdov, D. et al. 2019]. Cybersecurity is additionally identified with information move and preparing with correspondence conventions, and accordingly such interchanges must be scrambled, in spite of the fact that in most cases, it doesn't occur [Cai, Y. et al. 2019]. Cybersecurity of IoT frameworks utilising distributed computing is another difficult issue, since IoT gadgets can be associated through a cloud, giving cloud correspondence and information assortment. For this situation, insurance against distributed denial-of-service assaults turns into a significant component of the framework security [Hosseini, M., Berlin, R.R., Sha, L.A. 2017], while having humanoid–machine cooperation, there is a likely threat of meddling in such interchanges, prompting changes to robots. On the off chance that there is no encryption or verification instrument that controls such an interface, the framework is inclined to man-in-the-center assaults. In the following decade, it is normal that each house will have automatons, for example, family automatons and home colleagues in every day errands. They may contain mouthpieces, camcorder and sensing devices that will gather datasets, including individual data about house and even individuals' well-being status. Deficient consideration about insurance of this private data may bring about picking up a control under such help robot and getting to private information by an unapproved substance.

Web of Mechanical Things reveals another idea that is proposed to plot the converging of mechanical technology advances with IoT frameworks and distributed computing. As a result of this combination, the communication between IoT, distributed computing and mechanical technology research fields is growing quickly. Imaginative innovations and applications are taking into consideration more intelligent automaton frameworks that will ensure interoperability, constant ability and independent coordinated effort. Taking everything into consideration, such IoRT applications are insufficiently designed to completely shift into mechanical settings, as the principle tests mentioned in the comparing literature are done within scholarly research facilities. The requirement for participation among numerous robots, offering spaces to people, offers a key issue in keen situations. Multifunction automaton coordination is as yet confronting a few concerns in regard to agreement systems, oversee control and correspondence toward the two frameworks and different automatons, and also directed in a systematic way. A specific test additionally relies upon the absence of help for heterogeneous automaton arrangements. In multi-robot tasks, it is impressively hard to coordinate, arrange and facilitate the IoRT advances from various makers, which regularly utilise unique frameworks. In the examination area, a few upgrades have been done in empowering certain highlights in multifunction–automation applications; however, further upgrades fundamentally are required to the dynamic association of the robot makers themselves. Additionally, human–machine interactions are becoming highly common, which is especially important in a few scenarios, for example, clinics, cafés and administration regions. In human–robotic interface, savvy automatons should react to settled human signals [He, W., Li, Z., Chen, C.L.P. 2017; Islam, M.J., Sattar, J. 2017]. At the equivalent time, relating information preparing intends to arrive at the most extreme independence and security in human–robotic interface. Be that as it may, there are as yet numerous imperatives that frustrate human–robotic interface improvement. New typologies of human–robotic interface have been concentrated in the most recent years, for example, eye tracking [130], voice association [Hunter, P. 2019] and natural acknowledgment [Stower, R. 2019]; however, they despite everything should be examined, as the greater part of them have been broken down just in research labs [Del Campo, G. et al. 2018], and they are not broadly utilised at this point. With the turn of events and headway of the cyber-physical systems and 4th Industrial Rebellion, an ever increasing number of businesses are embracing far-off working answers for security reasons, where human administrators participate with automatons distantly, reconstructing and controlling them from a sheltered domain. In such situations, human–robotic interface accept another and more significant importance, as the requirement for progresses in innovation and applications can be very useful in permitting human administrators to distantly lead through modern automatons [Bukata, L. et al. 2016]. Headways in IoRT advances could be the response for another better approach to oversee modern activities. Of late, distant working is receiving a great deal of attention because of the major benefits that it brings. It has been indicated that off-site working prompts expanded execution in terms of creation, resulting in an improved work-life balance [Dragoni, N., Giaretta, A., Mazzara, M. 2018].

IoRT frameworks can further improve such progressions, opening additional opportunities identified with distantly working with modern robots. Moreover, a similar worldview can be effectively sent out to some other use of this study. For example, on account of training, understudies can encounter more advantages from the appropriation of social automatons masterminded in an IoRT situation. Distant training could be amazingly useful for a great deal of understudies, especially the ones influenced by illnesses that compel them to remain at home. Thus, leading to the certainty that collaboration among people and mechanical devices for instructive purposes should be developed [Khalid, A. et al. 2018]; furthermore, extending logical examinations on IoRT advancements will be absolutely critical later.

The development of the IoRT applications, notwithstanding the requirement for creating multifunctional automatons and human–robotic interface frameworks in savvy conditions, is prompting one of the primary issues concerning the period of 4th Industrial Revolution: vitality utilisation. Previously, energy utilisation was not a well studied subject, since passion for the study of energy utilisation has grown among research gathererings in just the last few decade. The trouble in assessing and improving the vitality effectiveness in shrewd conditions is principally because of the absence of the specific comprehension of vitality utilisation conduct [Wegner, A., Graham, J., Ribble, E. 2017]. To survive this issue, noteworthy consideration and endeavors must be devoted to accumulate data about vitality utilisation from shrewd sensors, utilising various philosophies [Wegner, A., Graham, J., Ribble, E. 2017]. For example, in the specific circumstance of proficient assembling, this data must be incorporated into the creation the board to accomplish economical procedures in the long haul. The handling of a lot of information in the IoRT applications prompts digital security issues. In such frameworks, it tends to be important to include outside computation and the IoRT's organisational capacity, such as automatons, at the edge, will most likely be unable to process and store enormous volumes of information. The primary digital security issues include shaky correspondence among clients and robots, confirmation issues, touchy information uncovering and frail default mechanical device design [Wegner, A., Graham, J., Ribble, E. 2017]. As security is key in organised automatons, new shrewd system designs are turning out to be fundamental to secure information data, yet even people in human–robotic interfacing-based frameworks. Indeed, security humanoid–machine-coordinated effort is an issue of basic significance in the modern condition, for the most part when it comes to IoRT frameworks [Wegner, A., Graham, J., Ribble, E. 2017], as digital assaults through the system or the Internet can bargain the smooth working of such mechanical frameworks. To tackle such issues, it is essential helping cyber-physical system security suppliers in recognising expected dangers by contrasting, breaking down, and gathering information from a few sources. Considering the time consumption of humanoid and personal computer to create IoRT frameworks, data with respect to creation and other mechanical control applications ought to be classified many more, profoundly checked, as appropriately gathering the data history could likewise be a key to guarantee digital-physical security. Assembling information must stay in a similar organisation, with a similar substance that it had upon its creation, and must be traded with some other reasonable mechanical framework, anyplace on the planet, utilising the equivalent secure convention. Digital-physical security issues can be overwhelmed by utilising incorporated confirmation and approval procedures to stream producing data [Wegner, A., Graham, J., Ribble, E. 2017], for example, information parts, appropriated duty, checked administrator control and granular approval. As the association among keen gadgets is fundamental to gain increasingly more ground in both mechanical and exploration fields, tackling issues identified with digital-physical security is of critical significance to improve the turn of events of IoRT frameworks into savvy spaces. Hence, security issues with respect to both cyber-physical systems and automaton association should be additionally examined.

2.4.1 Limitations on the Use of Smart Programs

1. There is no independent decision so far, robots have been a major aid. They support people with individual production steps or turn workers into questions. Wise conversations can grow from junior assistants to around work assistants with multiple technologies pioneers like Bill Gates, Elon Musk or Stephen Hawking; however, human intelligence will surpass AI within 15 years. If progress continues like this, it should be considered soon in the future; it will no longer be a decision maker, but a robot with its AI. When robots make important decisions and pass people in their careers, this can change for the better attitude toward Industry 4.0. In the state of development so far, however, people no longer have the full power to make decisions and that only auxiliary activities assigned robots, at least in the processing industry. The question is whether robots can make profitable decisions. Unlike most people, it is independent of the system that does not make its decisions based on the nature, however, on the basis of the policy process only. The robot announces the file a decision without feelings, so there will be a few misunderstandings in communication. However, it leaves the power to make decisions and people have a definite opportunity for social development acceptance of programs at establishment. What's right, however, the content is a decision.
2. There are no killings to kill. A no-go area for AI science research with intelligent weaponry programs is open fire without a personal decision made. In this context, radio-controlled weapons such as drones used by armed forces should be separated from the wise sensory systems that provide target support targeted selection and acquisition. Consequences of malfunction of such equipment are great, so all of them are highly desirable bans on private arm systems.
3. There is no use of heavy machinery or removal steps that do not happen now by engineering for technical reasons.

Another constraint in the field of robotics is manufacturing, and self-sufficient programs are now the state of the art. There are still complex steps of technical work that cannot be taken by a machine or its mechanical capture may not work properly. Composing can be considered one of the limitations of intellectual development programs. No standalone program can be set to generate novel ideas in the same way as state-of-the-art research did. Although needs can be analysed with great use of data, ideas for development are still in the people.

2.5 CONCLUDING REMARKS

The introduced composition sketched out those key innovations of 4th Industrial Rebellion that prompted the advancement of the IoRT frameworks. In particular, a review of how cyber-physical systems oversee collaboration between physical and virtual universes has been completed, planning to present the idea of IoRT. The designs of the two frameworks have been portrayed, and distinctive IoRT applications in various keen spaces have been sketched out, indicating how customary mechanical applications have profited by IoRT-constructed frameworks. The IoRT applications in keen spaces have been investigated, uncovering that IoRT frameworks are presently essential in various situations, beginning from the ones remembered for the modern field, for example, manufacturing and farming, to the ones that influence the regular day to day existence, for example, preventive medicine, training and scrutiny. IoRT applications can likewise establish a framework for the advancement of different areas that are outside the mechanical circle, for example, diversion, visits to exhibition halls, or sports rivalries, meaning to improve an ever increasing number of parts of human life. In addition, it has been seen that the improvement of IoRT frameworks can be the response to appropriately manage the need of distant working, where the new necessities of far-off transmission among people and automatons could be the response for additional fulfillment and efficiency. This overview has been for the most part centered around those IoRT-constructed frameworks in the mechanical and creative fields. IoRT issues and difficulties have been profoundly breaking down, indicating the expanding requirement for examination in machine-to-machine participation, most importantly in mixed systems of automatons, humanoid–machine interfaces for humanoid-fixed communications, vitality the board for the advancement of proficiency and digital security to ensure delicate information. The conceived headways in proficiency, vigor and security will open additional opportunities in even new areas of use, which will be advantage from the new spread of the most recent automated and system innovations.

REFERENCES

Adamo, F.; Attivissimo, F.; Di Nisio, A.; Carducci, C.G.C.; Spadavecchia, M.; Guagnano, A.; Goh, M. 2017, Comparison of Current Sensors for Power Consumption Assessment of Wireless Sensors Network Nodes. In Proceedings of the 2017 IEEE International Workshop on Measurement and Networking (M&N), Naples, Italy.

Afanasyev, I.; Mazzara, M.; Chakraborty, S.; Zhuchkov, N.; Maksatbek, A.; Kassab, M.; Distefano, S. 2020, Towards the Internet of Robotic Things: Analysis, Architecture, Components and Challenges. In Proceedings Sensors, 12th International Conference on Developments in E-Systems Engineering (DeSE), Kazan, Russia.

Alcácer, V.; Cruz-Machado, V. 2019, Scanning the Industry 4.0: A Literature Review on Technologies for Manufacturing Systems. Eng. Sci. Technol. Int. J. 22(3), 899–919.

Ali, M.H.; Aizat, K.; Yerkhan, K.; Zhandos, T.; Anuar, O. 2018, Vision-Based Robot Manipulator for Industrial Applications. Procedia Comput. Sci.

Amin, R.; Islam, S.H.; Biswas, G.; Khan, M.K.; Kumar, N. 2018, A Robust and Anonymous Patient Monitoring System Using Wireless Medical Sensor Networks. Future Gener. Comput. Syst. 80(C), 483–495.

Ashok Kumar, M.; Thirumurugan, T. 2018, Integrated IOT Based Design and Android Operated Multi-Purpose Field Surveillance Robot for Military Use. In Proceedings of the International Conference for Phoenixes on Emerging Current Trends in Engineering and Management (PECTEAM 2018), Tamil Nadu, India.

Basnet, B.; Bang, J. 2018, The State-of-the-Art of Knowledge-Intensive Agriculture: A Review on Applied Sensing Systems and Data Analytics. J. Sens. 2018, 1–14.

Batth, R.S.; Nayyar, A.; Nagpal, A., 2018, Internet of Robotic Things: Driving Intelligent Robotics of Future-Concept, Architecture, Applications and Technologies. In Proceedings of the 2018 4th International Conference on Computing Sciences (ICCS), Jalandhar, India.

Belbachir, A.; Benabid, S. 2016, Smart Communication for Cooperative Wireless Sensor Network. In Proceedings of the 2016 International Conference on Applied Electronics (AE), Pilsen, Czech Republic.

Blaser, W.J.; Oppong, J.; Hart, S.P.; Landolt, J.; Yeboah, E.; Six, J. 2018, Climate-Smart Sustainable Agriculture in Low-to-Intermediate Shade Agroforests. Nat. Sustain. 1(5), 234–239.

Bonaci, T.; Herron, J.; Yusuf, T.; Yan, J.; Kohno, T.; Chizeck, H.J. 2015, To Make a Robot Secure: An Experimental Analysis of Cyber Security Threats against Teleoperated Surgical Robots.

Bukata, L.; Sucha, P.; Hanzálek, Z.; Burget, P. 2016, Energy Optimization of Robotic Cells. IEEE Trans. Ind. Inform.2016, 13, 92–102.

Cai, Y.; Tang, W.; Zhang, B.; Wang, Y. 2019, Collaborative Configuration for Distributed Energy Storages and Cyber Systems in Low-Voltage Distribution Networks with High Penetration of PV Systems. IET Cyber Phys. Syst. Theory Theor. Appl.

Campbell, B.M.; Thornton, P.; Zougmoré, R.; Van Asten, P.; Lipper, L. 2014, Lipper, L. Sustainable Intensification: What Is Its Role in Climate Smart Agriculture? Curr. Opin. Environ. Sustain. 8, 39–43.

Cao, H.; Huang, X.; Zhuang, J.; Xu, J.; Shao, Z. C. 2018, IoT-Robot Cloud and IoT Assisted Indoor Robot for Medicine Delivery. In Proceedings of the 2018 Joint International Advanced Engineering and Technology Research Conference (JIAET 2018), Xi'an, China.

Cecil, J.; Albuhamood, S.; Ramanathan, P.; Gupta, A. 2019, An Internet-of-Things (IoT) Based Cyber Manufacturing Framework for the Assembly of Micro Devices. Int. J. Compute. Integr. Manuf. 32, 430–440.

Chang, W.; Lin, S.; Hsu, J.; Hsu, B. 2019, Automatic Path Planning of Robot for Intelligent Manufacturing Based on Network Remoted Controlling and Simulation. In Proceedings of the 2019 4th Asia-Pacific Conference on Intelligent Robot Systems (ACIRS), Nagoya, Japan.

Chen, F.; Selvaggio, M.; Caldwell, D.G.. 2019, Dexterous Grasping by Manipulability Selection for Mobile Manipulator with Visual Guidance. IEEE Trans. Ind. Inform. 15(2), 1202–1210.

Chen, M.; Zhou, J.; Tao, G.; Yang, J.; Hu, L. 2018, Wearable Affective Robot. IEEE Access. 6(2018), 64766–64776.

Chen, X.; Qin, S. 2017, Approach to High Efficient Hierarchical Pathfinding of Indoor Mobile Service Robots Based on Grid Map and Floyd-Warshall Algorithm. In Proceedings of the 2017 Chinese Automation Congress (CAC), Jinan, China, 20–22 October 2017; IEEE: Piscataway, NJ, USA.

Chukwuemeka, C.; Habib, M. 2018, Development of Autonomous Networked Robots (ANR) for Surveillance: Conceptual Design and Requirements. In Proceedings of the IECON 2018—44th Annual Conference of the IEEE Industrial Electronics Society, Washington.

Coronado, P.D.U.; Ahuett-Garza, H.; Morales-Menendez, R.; Castañón, P.O.; Dávila, L.D.;
    Escalera, M.R.F.; Escalera, M.R.F. 2017, Connectivity of a Modular Electric Vehicle by
    the Use of a Mobile Device. Adv. Mech. Eng. 9(7), 1–9.
Del Campo, G.; Calatrava, S.; Cañada, G.; Olloqui, J.; Martinez, R.; Santamaria, A. 2018,
    Santamaria, A., IoT Solution for Energy Optimization in Industry 4.0: Issues of a Real-
    Life Implementation. In Proceedings of the 2018 Global Internet of Things Summit
    (GIoTS), Bilbao, Spain.
Dharmasena, T.; de Silva, R.; Abhayasingha, N.; Abeygunawardhana, P. 2019, Autonomous
    Cloud Robotic System for Smart Agriculture. In Proceedings of the 2019 Moratuwa
    Engineering Research Conference (MERCon), Moratuwa, Sri Lanka.
Dragoni, N.; Giaretta, A.; Mazzara, M. 2018, The Internet of Hackable Things. In Proceedings
    of the 5th International Conference in Software Engineering for Defence Applications,
    Rome, Italy, 10 May 2016; Springer International Publishing: Cham, Switzerland.
Drozdov, D.; Patil, S.; Dubinin, V.; Vyatkin, V. 2019, Towards Formal ASM Semantics of
    Timed Control Systems for Industrial CPS. In Proceedings of the 2019 24th IEEE
    International Conference on Emerging Technologies and Factory Automation (ETFA),
    Zaragoza, Spain.
Duckett, T.; Pearson, S.; Blackmore, S.; Grieve, B. 2018, Agricultural Robotics: The Future of
    Robotic Agriculture. arXiv 2018, ar Xiv:1806.06762. Sensors 2020.
Faryadi, S.; Davoodi, M.; Mohammadpour Velni, J. 2019, Agricultural Field Coverage Using
    Cooperating Unmanned Ground Vehicles. In Proceedings of the Dynamic Systems and
    Control Conference, American Society of Mechanical Engineers, Park City, UT, USA.
Fenning, R.M.; Baker, J.K.; Baucom, B.R.; Erath, S.A..; Howland, M.A.; Moffitt, J. 2017;
    Howland, M.A.; Moffitt, J., Electrodermal Variability and Symptom Severity in
    Children with Autism Spectrum Disorder.
Fierro, R.; Chaimowicz, L.; Kumar, V. 2018, Multi-Robot Cooperation. In Autonomous
    Mobile Robots; CRC Press; Boca Raton, FL, USA
Guggero, M.; Di Paola, D.; Petitti, A.; Caviglione, L. 2019, When Time Matters: Predictive
    Mission Planning in Cyber-Physical Scenarios. IEEE Access. 7, 11246–11257.
Garcia, C.A.; Montalvo-Lopez, W.; Garcia, M.V. 2020, Human-Robot Collaboration Based on
    Cyber-Physical Production System and MQTT. Procedia Manufacturing. 42, 315–321.
Garcia, M. A. R.; Rojas, R.; Gualtieri, L.; Rauch, E.; Matt, D. 2019, A Human-in-the-Loop
    Cyber-Physical System for Collaborative Assembly in Smart Manufacturing. Procedia
    CIRP, 81, 600–605.
García-Tejero, I.F.; Carbonell, R.; Ordoñez, R.; Torres, F.P.; Zuazo, V.H.D. 2020, Conservation
    Agriculture Practices to Improve the Soil Water Management and Soil Carbon
    Storage in Mediterranean Rainfed Agro-Ecosystems. In Soil Health Restoration and
    Management; Springer: Berlin/Heidelberg, Germany.
Grau, A.; Indri, M.; Bello, L.L.; Sauter, T. 2017, Industrial Robotics in Factory Automation:
    From the Early Stage to the Internet of Things. In Proceedings of the IECON 2017—
    43rd Annual Conference of the IEEE Industrial Electronics Society, Beijing, China.
Grieves, M.; Vickers, J. 2017, Digital Twin: Mitigating Unpredictable, Undesirable Emergent
    Behavior in Complex Systems. In Transdisciplinary Perspectives on Complex Systems:
    New Findings and Approaches; Kahlen, F.J., Flumerfelt, S., Alves, A., Eds.; Springer
    International Publishing: Cham, Switzerland.
Grigorescu, S.D.; Argatu, F.C.; Paturca, S.V.; Cepisca, C.; Seritan, G.C.; Adochiei, F.C.;
    Enache, B. 2019, Robotic Platform with Medical Applications in the Smart City
    Environment. In Proceedings of the 2019 11th International Symposium on Advanced
    Topics in Electrical Engineering (ATEE), Bucharest, Romania.
Halme, R.J.; Lanz, M.; Kämäräinen, J.; Pieters, R.; Latokartano, J.; Hietanen, A. 2018,
    Review of Vision-Based Safety Systems for Human-Robot Collaboration. Procedia
    CIRP. 72, 111–116.

He, W.; Li, Z.; Chen, C.L.P. 2017, A Survey of Human-Centered Intelligent Robots: Issues and Challenges. IEEE/CAA J. Autom. Sin. 4(4), 602–609.

Hosseini, M.; Berlin, R.R.; Sha, L.A. 2017, Physiology-Aware Communication Architecture for Distributed Emergency Medical CPS. In Proceedings of the 8th International Conference on Cyber-Physical Systems, ICCPS '17, Pittsburgh, PA; Association for Computing Machinery: New York, NY, USA.

Hunter, P. 2019, Remote Working in Research: An Increasing Usage of Flexible Work Arrangements Can Improve Productivity and Creativity. EMBO Rep. 20(1), e47435.

Islam, M.J.; Sattar, J. 2017, Mixed-Domain Biological Motion Tracking for Underwater Human- Robot Interaction. In Proceedings of the 2017 IEEE International Conference on Robotics and Automation (ICRA), Singapore.

Jawad, H.M.; Nordin, R.; Gharghan, S.K.; Jawad, A.M.; Ismail, M. 2017, Energy- Efficient Wireless Sensor Networks for Precision Agriculture: A Review. Sensors. 17(8), 1781.

Jiang, J.R. 2018, An Improved Cyber-Physical Systems Architecture for Industry 4.0 Smart Factories. Adv. Mech. Eng. 10(6).

Ju, C.; Son, H.I. 2020, Modeling and Control of Heterogeneous Agricultural Field Robots Based on Ramadge–Wonham Theory. IEEE Robot. Autom. Lett., 5, 48–55.

Kalpana, M.A.T.M.; Tyagi, M.A.. 2017, Bellman Ford Shortest Path Algorithm Using Global Positioning System. Int. Res. J. Eng. Technol. 4(04), 12.

Ke, Q.; Xiaogang, L. 2016, Internet-of-Things Monitoring System of Robot Welding Based on Software Defined Networking. In Proceedings of the 2016 First IEEE International Conference on Computer Communication and the Internet (ICCCI), Wuhan, China, 13–15.

Kehoe, B.; Patil, S.; Abbeel, P.; Goldberg, K.A. 2015, Survey of Research on Cloud Robotics and Automation. IEEE Trans. Autom. Sci. Eng. 12(2), 398–409.

Khakimov, A.; Muthanna, A.; Kirichek, R.; Koucheryavy, A.; Muthanna, M.S.A. 2017, Investigation of Methods for Remote Control IoT-Devices Based on Cloud Platforms and Different Interaction Protocols. In Proceedings of the 2017 IEEE Conference of Russian Young Researchers in Electrical and Electronic Engineering (EIConRus), St. Petersburg, Russia.

Khalid, A.; Kirisci, P.; Khan, Z.H.; Ghrairi, Z.; Thoben, K.D.; Pannek, J. 2018, Security Framework for Industrial Collaborative Robotic Cyber-Physical Systems. Comput. Ind. 97, 132–145.

Khatri-Chhetri, A.; Aggarwal, P.K.; Joshi, P.K.; Vyas, S. 2017, Farmers' Prioritization of Climate-Smart Agriculture (CSA) Technologies. Agric. Syst. 151, 184–191.

Koken, B. 2015, Cloud Robotics Platforms. Interdiscip. Descr. Complex Syst. INDECS. 13(1), 26–33.

Kondratenko, Y.P. 2015, Robotics, Automation and Information Systems: Future Perspectives and Correlation with Culture, Sport and Life Science. In Decision Making and Knowledge Decision Support Systems: VIII International Conference of RACEF, Barcelona, Spain, November 2013 and International Conference MS 2013, Chania Crete, b Greece, November 2013; Gil-Lafuente, A.M., Zopounidis, C., Eds.; Springer International Publishing: Cham, Switzerland.

Krishna, K.L.; Silver, O.; Malende, W.F.; Anuradha, K. 2017, Internet of Things Application for Implementation of Smart Agriculture System. In Proceedings of the 2017 International Conference on I-SMAC (IoT in Social, Mobile, Analytics and Cloud), Palladam.

Kumar, A. 2018, Methods and Materials for Smart Manufacturing: Additive Manufacturing, Internet of Things, Flexible Sensors and Soft Robotics. Industry 4.0 and Smart Manufacturing. Manuf. Lett. 15, 122–125.

Lee, Y.K.; Goh, Y.H.; Tew, Y. 2018, Cyber Physical Autonomous Mobile Robot (CPAMR) Framework in the Context of Industry 4.0. MATEC Web Conf. 167, 02005. EDP Sciences.

Leitão, P.; Karnouskos, S.; Ribeiro, L.; Lee, J.; Strasser, T.; Colombo, A.W. 2016, Smart Agents in Industrial Cyber–Physical Systems. In Proc. IEEE 20. 104(5), 1086–1101.

Li, T.S.; Liu, C.; Kuo, P.; Fang, N.; Li, C.; Cheng, C.; Hsieh, C.; Wu, L.; Liang, J.; Chen, C. 2017, A Three-Dimensional Adaptive PSO-Based Packing Algorithm for an IoT-Based Automated e-Fulfillment Packaging System. IEEE Access. 5, 9188–9205.

Li, X.; Ma, Z.; Chu, X.; Liu, Y. 2019, A Cloud-Assisted Region Monitoring Strategy of Mobile Robot in Smart Greenhouse. Mob. Inf. Syst.

Liu, Q.; Hua, P.; Sultan, A.; Shen, L.; Mueller, E.; Boerner, F. 2020, Study of the Integration of Robot In Cyber-Physical Production Systems. In Proceedings of the 2019 International Conference on Cyber-Enabled Distributed Computing and Knowledge Discovery (CyberC), Guilin, China.

Liu, Q.; Liu, Z.; Xu, W.; Tang, Q.; Zhou, Z.; Pham, D.T. 2019, Human-Robot Collaboration in Disassembly for Sustainable Manufacturing. Int. J. Prod. Res. 57(12), 4027–4044.

Liu, Y.; Peng, Y.; Wang, B.; Yao, S.; Liu, Z. 2017, Review on Cyber-Physical Systems. IEEE/ CAA J. Autom. Sin. 4(1), 27–40.

Lu, Y. 2017, Industry 4.0: A Survey on Technologies, Applications and Open Research Issues. J. Ind. Inf. Integr. 6, 1–10.

Madhira, K.; Mehta, S.; Bollineni, R.; Kavathia, D. 2017, AGWallP—Automatic Guided Wall Painting System. In Proceedings of the 2017 Nirma University International Conference on Engineering (NUiCONE), Ahmedabad, India.

Mahbub, M.A. 2020, A Smart Farming Concept Based on Smart Embedded Electronics, Internet of Things and Wireless Sensor Network. Internet Things 2020, 9, 100161.

Mat, I.; Kassim, M.R.M.; Harun, A.N.; Yusoff, I.M. 2018, Smart Agriculture Using Internet of Things. In Proceedings of the 2018 IEEE Conference on Open Systems (ICOS), Langkawi, Malaysia.

Mazzara, M.; Afanasyev, I.; Sarangi, S.; Distefano, S.; Kumar, V. 2019, A Reference Architecture for Smart and Software-Defined Buildings. In Proceedings of the 2019 IEEE International Conference on Smart Computing (SMARTCOMP), Washington, DC, USA.

McConnell, M.D.; Burger, L.W. Jr. 2017, Precision Conservation to Enhance Wildlife Benefits in Agricultural Landscapes. Precis. Conserv. Geospat. Tech. Agric. Nat. Resour. Conserv. 2017, 59, 285–312.

Meghana, S.; Nikhil, T.V.; Murali, R.; Sanjana, S.; Vidhya, R.; Mohammed, K.J. 2017, Design and Implementation of Surveillance Robot for Outdoor Security. In Proceedings of the 2017 2nd IEEE International Conference on Recent Trends in Electronics, Information Communication Technology (RTEICT), Bangalore, India.

Mekala, M.S.; Viswanathan, P. 2017, A Survey: Smart Agriculture IoT with Cloud Computing. In Proceedings of the 2017 International Conference on Microelectronic Devices, Circuits and Systems (ICMDCS), Vellore, India, 10–12 August 2017; IEEE: Piscataway, NJ, USA.

Micoli, G.; Boccadoro, P.; Valecce, G.; Petitti, A.; Colella, R.; Milella, A. ; Grieco, L.A.; Grieco, L.A. 2019, ASAP: A Decentralized Slot Reservation Policy for Dynamic 6TiSCH Networks in Industrial IoT. In Proceedings of the 2019 IEEE International Conference on Communications Workshops (ICC Workshops), Shanghai, China.

Milella, A.; Marani, R.; Petitti, A.; Reina, G. 2019, In-Field High Throughput Grapevine Phenotyping with a Consumer-Grade Depth Camera. Comput. Electron. Agric. 156, 293–306.

Mosavi, A.; Varkonyi-Koczy, A.R. 2017, Integration of Machine Learning and Optimization for Robot Learning. In Recent Global Research and Education: Technological Challenges; Jablonski, R., Szewczyk, R., Eds.; Springer International Publishing: Cham, Switzerland.

Mrudul, K.; Mandava, R.K.; Vundavilli, P.R. 2018, An Efficient Path Planning Algorithm for Biped Robot Using Fast Marching Method. Procedia Comput. Sci. 133, 116–123.

Nathalie, M.; David, S.R. 2013, Wireless Sensor and Robot Networks: From Topology Control to Communication Aspects; World Scientific: Singapore.

Nayyar, A.; Puri, V.; Nguyen, N.G.; Le, D.N. 2018, Smart Surveillance Robot for Real-Time Monitoring and Control System in Environment and Industrial Applications. In Information Systems Design and Intelligent Applications; Bhateja, V., Nguyen, B.L., Nguyen, N.G., Satapathy, S.C., Le, D.N., Eds.; Springer: Singapore.

Pai, V.; Shenoy, U.K.K. 2019, LowPan—Performance Analysis on Low Power Networks. In International Conference on Computer Networks and Communication Technologies; Smys, S., Bestak, R., Chen, J.I.Z., Kotuliak, I., Eds.; Springer: Singapore.

Pedersen, M.R.; Nalpantidis, L.; Andersen, R.S.; Schou, C.; Bøgh, S.; Krüger, V.; Madsen, O. 2016, Robot Skills for Manufacturing: From Concept to Industrial Deployment. Robot. Comput. Integr. Manuf. 37, 282–291.

Petitti, A.; Di Paola, D.; Milella, A.; Lorusso, A.; Colella, R.; Attolico, G.; Caccia, M.A. 2016, Network of Stationary Sensors and Mobile Robots for Distributed Ambient Intelligence. IEEE Intell. Syst. 31(6), 28–34.

Prasad, D.; Singla, K.; Baggan, V. 2019; others. System Model for Smart Precision Farming for High Crop Yielding. J. Comput. Theor. Nanosci. 16(10), 4406–4411.

Qing, G.; Zheng, Z.; Yue, X. 2017, Path-Planning of Automated Guided Vehicle Based on Improved Dijkstra Algorithm. In Proceedings of the 2017 29th Chinese Control and Decision Conference (CCDC), Chongqing, China, 28–30 May 2017; IEEE: Piscataway, NJ.

Ramson, S.J.; Moni, D.J. 2017, Applications of Wireless Sensor Networks—A Survey. In Proceedings of the 2017 International Conference on Innovations in Electrical, Electronics, Instrumentation and Media Technology (ICEEIMT), Coimbatore, India, 3–4 February 2017; IEEE: Piscataway, NJ.

Razafimandimby, C.; Loscri, V.; Vegni, A.M. 2020, Towards Efficient Deployment in Internet of Robotic Things. In Integration, Interconnection, and Interoperability of IoT Systems; Gravina, R., Palau, C.E., Manso, M., Liotta, A., Fortino, G., Eds.; Springer International Publishing: Cham, Switzerland.

Reina, G.; Milella, A.; Rouveure, R.; Nielsen, M.; Worst, R. ; Blas, M.R.; Blas, M.R., 2016, Ambient Awareness for Agricultural Robotic Vehicles. Biosyst. Eng. 146, 114–132.

Romeo, L.; Petitti, A.; Colella, R.; Valecce, G.; Boccadoro, P.; Milella, A.; Grieco, L.A. 2020, Automated Deployment of IoT Networks in Outdoor Scenarios Using an Unmanned Ground Vehicle. In Proceedings of the 21st IEEE 2020 International Conference on Industrial Technology (ICIT2020), Buenos Aires, Argentina.

Romero-Hdz, J.; Saha, B.; Toledo-Ramirez, G.; Beltran-Bqz, D. 2016, Welding Sequence Optimization Using Artificial Intelligence Techniques: An Overview. SSRG Int. J. Comput. Sci. Eng. 3, 90–95.

Roopaei, M.; Rad, P.; Choo, K.K.R. 2017, Cloud of Things in Smart Agriculture: Intelligent Irrigation Monitoring by Thermal Imaging. IEEE Cloud Comput. 4(1), 10–15.

Rosenfeld, A.; Agmon, N.; Maksimov, O.; Kraus, S. 2017, Intelligent Agent Supporting Human–Multi-Robot Team Collaboration. Artif. Intell. 252, 211–231.

Saad, W.H.M.; Karim, S.A.A.; Azhar, N.; Manap, Z.; Soon, Y.Y.; Ibrahim, M.M.; Ibrahim, M.M. 2018, Line Follower Mobile Robot for Surveillance Camera Monitoring System. J. Telecomm. Electron. Compute. Eng. (JTEC). 10(2–7), 1–5.

Saponara, S.; Pilato, L.; Fanucci, L. 2016, Exploiting CCTV Camera System for Advanced Passenger Services On-Board Trains. In Proceedings of the 2016 IEEE International Smart Cities Conference (ISC2). Manchester, United Kingdom, 1–6. IEEE.

Saraswathi, M.; Murali, G.B.; Deepak, B. 2018, Optimal Path Planning of Mobile Robot Using Hybrid Cuckoo Search-Bat Algorithm. Procedia Comput. Sci. 133, 510–517.

Saravanan, D.; Archana, G.; Parthiban, R. 2018, Voice Controlled Humanoid Robotic Car for Smart Agriculture Using Arduino and Android Smart Watch. Int. J. Pure Appl. Math. 119, 829–833.

Serebrenny, V.; Shereuzhev, M.; Metasov, I. 2018, Approaches to the Robotization of Agricultural Mobile Machines. MATEC Web Conf. EDP Sci. 161, 03014.

Scilimati, V.; Petitti, A.; Boccadoro, P.; Colella, R.; Di Paola, D.; Milella, A..; Alfredo Grieco, L.; Alfredo Grieco, L. 2017, Industrial Internet of Things at Work: A Case Study Analysis in Robotic-Aided Environmental Monitoring. IET Wirel. Sens. Syst. 7(5), 155–162.

Shah, M.S.; Borole, P.B. 2016, Surveillance and Rescue Robot Using Android Smartphone and the Internet. In Proceedings of the 2016 International Conference on Communication and Signal Processing (ICCSP), Melmaruvathur, India.

Shih, C.S.; Chou, J.J.; Reijers, N.; Kuo, T.W. 2016, Designing CPS/IoT Applications for Smart Buildings and Cities. IET Cyber Phys. Syst. Theory Theor. Appl. 1(1), 3–12.

Singh, R.; Samkaria, R.; Gehlot, A.; Choudhary, S. 2018, Design and Development of IoT Enable Multi Robot System for Search and Rescue Mission. IJWA. 10(2), 51–63.

Stower, R. 2019, The Role of Trust and Social Behaviors in Children's Learning from Social Robots. In Proceedings of the 2019 8th International Conference on Affective Computing and Intelligent Interaction Workshops and Demos (ACIIW), Cambridge, UK.

Tomiyama, T.; Moyen, F. 2018, Cranfield University, Cranfield MK43 0AL, UK, Resilient Architecture for Cyber-Physical Production Systems. CIRP Ann. 67(1), 161–164.

Tyagi, N.K.; Joshi, P.K. 2019, Index-Based Insurance for Mitigating Flood Risks in Agriculture: Status, Challenges and Way Forward. In Climate Smart Agriculture in South Asia: Technologies, Policies and Institutions; Pal, B.D., Kishore, A., Joshi, P.K., Tyagi, N.K., Eds.; Springer: Singapore.

Uddin, M.A.; Mansour, A.; Le Jeune, D.; Aggoune, E.H.M. 2017, Agriculture Internet of Things: AG-IoT. In Proceedings of the 2017 27th International Telecommunication Networks and Applications Conference (ITNAC), Melbourne, Australia.

Valecce, G.; Micoli, G.; Boccadoro, P.; Petitti, A.; Colella, R.; Milella, A.; Alfredo Grieco, L. 2019; Alfredo Grieco, L., Robotic-Aided IoT: Automated Deployment of a 6TiSCH Network Using an UGV. IET Wirel. Sens. Syst. 9(6), 438–446.

van Erp, T.; Seliger, G. 2016, Opportunities of Sustainable Manufacturing in Industry 4.0. 13th Global Conference on Sustainable Manufacturing—Decoupling Growth from Resource Use, Procedia CIRP. 40, 536–541.

Vermesan, O.; Bröring, A.; Tragos, E.; Serrano, M.; Bacciu, D.; Chessa, S.; Gallicchio, C.; Micheli, A.; Dragone, M.; Saffiotti, A.; et al. 2017, Internet of Robotic Things: Converging Sensing/Actuating, Hypo Connectivity, Artificial Intelligence and IoT Platforms. In Cognitive Hyper Connected Digital Transformation: Internet of Things Intelligence Evolution; River Publishers: Gistrup, Denmark.

Vos, M.; Cranefield, J. 2017, RFID/Internet of Things Systems on the Boundary between Public and Private Sectors: An Ant Study of Multiplicity. In Proceedings of the 25th European Conference on Information Systems (ECIS), Guimarães, Portugal.

Wan, J.; Tang, S.; Shu, Z.; Li, D.; Wang, S.; Imran, M.; Vasilakos, A.V. 2016, Software-Defined Industrial Internet of Things in the Context of Industry 4.0. IEEE Sensors Journal, 1. 16(20), 7373–7380.

Wang, S.; Wan, J.; Imran, M.; Li, D.; Zhang, C. 2018, Cloud-Based Smart Manufacturing for Personalized Candy Packing Application. J. Supercomput. 74(9), 4339–4357.

Wegner, A.; Graham, J.; Ribble, E. 2017, A New Approach to Cyberphysical Security in Industry 4.0. In Cybersecurity for Industry 4.0: Analysis for Design and Manufacturing; Thames, L., Schaefer, D., Eds.; Springer International Publishing: Cham, Switzerland.

Wolfert, S.; Ge, L.; Verdouw, C.; Bogaardt, M.J. 2017, Big Data in Smart Farming—A Review. Agric. Syst.

Wollschlaeger, M.; Sauter, T.; Jasperneite, J. 2017, The Future of Industrial Communication: Automation Networks in the Era of the Internet of Things and Industry 4.0. IEEE Ind. Electron. Mag. 11(1), 17–27.

Xin, L.; Xiangyuan, H.; Ziqi, Y.; Xiaoning, Q.; Yingkui, D. 2018, The Algebraic Algorithm for Path Planning Problem of AGV in Flexible Manufacturing System. In Proceedings of the 2018 37th Chinese Control Conference (CCC), Wuhan, China, Piscataway, NJ.

Xu, H.; Yu, W.; Griffith, D.; Golmie, N. 2018, A Survey on Industrial Internet of Things: A Cyber-Physical Systems Perspective. IEEE Access. 6, 78238–78259.

Xu, X.; Li, X.; Zhang, R. 2019, Remote Configurable Image Acquisition Lifting Robot for Smart Agriculture. In Proceedings of the 2019 IEEE 4th Advanced Information Technology, Electronic and Automation Control Conference (IAEAC), Chengdu, China.

Yao, B.; Zhou, Z.; Wang, L.; Xu, W.; Liu, Q.; Liu, A. 2018, Sensorless and Adaptive Admittance Control of Industrial Robot in Physical Human- Robot Interaction. Robot. Comput. Integr. Manuf. 51, 158–168.

Yao, X.; Zhou, J.; Lin, Y.; Li, Y.; Yu, H.; Liu, Y. 2019, Smart Manufacturing Based on Cyber-Physical Systems and Beyond.

Yin, S.; Li, X.; Gao, H.; Kaynak, O. 2015, Data-Based Techniques Focused on Modern Industry: An Overview. IEEE Trans. Ind. Electron. 62(1), 657–667.

Yu, M.; Zhu, M.; Chen, G.; Li, J.; Zhou, Z. 2016, A Cyber-Physical Architecture for Industry 4.0-Based Power Equipment's Detection System. In Proceedings of the 2016 International Conference on Condition Monitoring and Diagnosis (CMD), Xi'an, China.

Zaman, S.; Comba, L.; Biglia, A.; Aimonino, D.R.; Barge, P.; Gay, P. 2019, Cost-Effective Visual Odometry System for Vehicle Motion Control in Agricultural Environments. Compute. Electron. Agric. 162, 82–94.

Zheng, P.; Wang, H.; Sang, Z.; Zhong, R.Y.; Liu, Y.; Liu, C.; Mubarok, K.; Yu, S.; Xu, X. 2018, Smart Manufacturing Systems for Industry 4.0: Conceptual Framework, Scenarios, and Future Perspectives. Front. Mech. Eng. 13(2), 137–150.

Zhou, H.; Lv, H.; Pang, Z.; Huang, X.; Yang, H.; Yang, G. 2019, IoT-Enabled Dual-Arm Motion Capture and Mapping for Telerobotics in Home Care. IEEE J. Biomed. Health Inform. 24(6), 1541–1549.

Zhou, K.; Liu, T.; Zhou, L. 2015, Industry 4.0: Towards Future Industrial Opportunities and Challenges. In Proceedings of the 12th International Conference on Fuzzy Systems and Knowledge Discovery (FSKD), Zhangjiajie, China, 15–17 August 2015.

Zimmerman, M.; Marvel, J.A. 2017, Smart Manufacturing and the Promotion of Artificially-Intelligent Human-Robot Collaborations in Small- and Medium-Sized Enterprises. In Proceedings of the Association for the Advancement of Artificial Intelligence, Arlington, MA.

3 The Role of Sensing Techniques in Precision Agriculture

Upendar K., Agrawal K.N. and Vinod Kumar S.

CONTENTS

3.1   INTRODUCTION

The Fourth Industrial Revolution, known as Industry 4.0, automates conventional manufacturing and industrial practices by employing modern smart technology. Large-scale machine-to-machine communication intelligence systems and the Internet of Things (IoT) are combined for improved communication, increased automation and self-monitoring and the production of smart machines that can analyse and interpret issues in the absence of human involvement. India is mainly an agrarian country, and almost 70% of its population directly or indirectly depend on agriculture for their survival. Agriculture is considered as a primary sector, and it is the basis for the development of secondary as well tertiary sectors. In the existing farming practices, farmers often visit agriculture sites during crop life to thoroughly understand the crop conditions. Due to this, approximately 70% of farmers' valuable time is employed only to observe and understand the present crop status instead of doing actual fieldwork (Navulur and Prasad, 2017). It is estimated by the Food and Agriculture Organization (FAO) that the global population is going to be approximately 9.6 billion in 2050. Therefore, there is a big challenge to produce 70% more food to feed such a huge, growing population.

After World War II, to meet the food demand of the country, the Green Revolution was introduced. The population was increasing, and, on the other side,

DOI: 10.1201/9781003122401-3

cultivable land availability for food production was not sufficient to meet the growing population. Hence, the Green Revolution in the year 1960 took place with an objective to increase productivity from a hectare of land. As part of the Green Revolution high-yielding seed varieties, irrigation facilities, pesticides, fertilisers and farm mechanisation were introduced. Due to the Green Revolution, farm productivity improved tremendously. But on the other side, there was a huge stress on the soil and serious environmental pollution due to excessive application of chemical fertilisers and pesticides. The World Health Organization (WHO) states that the continued, concentrated and indiscriminate agrochemical usage negatively affected soil biodiversity, causing long-term adverse effects on human and animal health. Every year chemical consumption is increasing: in the year 1958, the chemical consumption was 500 Mt, but in 2019, it was found to be 2,167,03 Mt (Anonymous, 2020a). In Indian agriculture, there is a labor shortage due to agriculture labor migrating from villages to urban areas, which is a major concern. Present agriculture practice and dumping industry wastages in the arable lands are causing serious soil poisoning worldwide. A solution to these two major issues is possible by employing engineering in agriculture. Adoption of sensors and mechanisation in the current agricultural practices bring better input use efficiency and timely field operations.

The second challenge is that new innovation should be environmentally friendly with no damage to the ecosystem and no impact on the climate.

Precision agriculture involves the use of cutting-edge technologies at the field level to precisely regulate agriculture inputs to be applied based on the status of the crop, soil and weather conditions. The current agricultural practices need to be modified to accommodate management decisions of site-specific management. Yield loss during crop production from various sources is shown in Figure 3.1 (Anonymous, 2020a). Sensors help in the early identification of plant health status so that suitable measures can be implemented immediately before the problem becomes too serious. Site-specific crop management practices are possible with the employment of sensors for variability detection in the field. Moreover, drones (unmanned aerial vehicles [UAVs]) for spraying and hyperspectral imaging and robots for fruit harvesting also play a key role in precision agriculture (Adão et al., 2017). Brown et al. (2008) compared the surface water runoff and ground deposits resulting from conventional broadcast spraying versus target spraying in dormant orchards. From the results it can be concluded

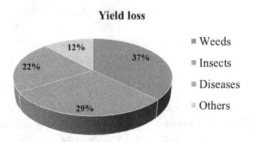

FIGURE 3.1   Yield loss due to pests (Anonymous, 2020a).

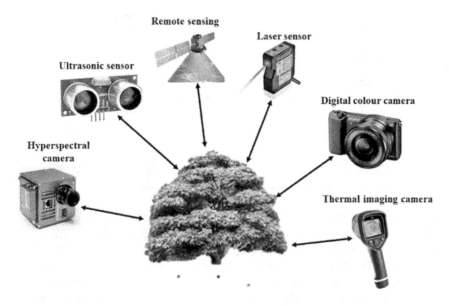

FIGURE 3.2 Various sensors for target detection and analysis.

that by stepping toward target spraying over conventional broadcast spraying, there was a drastic reduction in ground deposition by 41% and in surface runoff by 44%. Various ground-based sensing technologies, such as machine vision, spectral analysis and distance-based sensors used in target detection, are shown in Figure 3.2. The role of imaging sensors in the various domains of agriculture has been increasing because computer vision and image processing applications in agriculture are increasing. Imaging sensors are used in various fields such as weed and main crop detection, plant stress detection, leaf disease detection and canopy volume estimation.

3.2 APPLICATION OF SENSORS IN PRECISION AGRICULTURE

3.2.1 SENSORS IN FERTILISER APPLICATION

Fertiliser is a chemical substance that supplies essential nutrients for plants' growth and fertility. The fertiliser dose to be applied depends on existing soil nutrients, crop health, soil type and the previous crop grown. Excessive use of nutrients can be incredibly harmful to plant health and can lead to environmental pollution due to soil poisoning. The excessive use of chemical fertilisers increases the cost of cultivation for farmers also. It is reported that crops absorb less than 50% nitrogen applied in the form of fertiliser, the remaining nitrogen stays in the atmosphere, or reaches the water bodies through runoffs. Using conventional methods for the estimation of soil nutrient levels is time-consuming, and it incurs more cost. Suitable sensors for nutrient status measurement of the crop in real time are required. There are many methods employed by researchers to assess the nutrient status of the crop. Using near-infrared (NIR) and infrared (IR) wavelength, the normalised difference

TABLE 3.1

NDVI Values of Plants at Different Conditions

Plant Status	NDVI Range
Dead plants or inanimate objects	0
Unhealthy plant	0–0.33
Moderately healthy plant	0.33–0.66
Very health plant	0.66–1

Source: Liu et al., 2018.

vegetation index (NDVI) was proposed by Liu et al. (2018). This vegetation index helps in estimating the present nitrogen status of the crop in real time. Green seeker is a handheld device that provides crop health status by measuring its NDVI value (Table 3.1). The NDVI value ranges from 0 to 1. Zero indicates that a plant is an unhealthy plant, whereas one indicates a healthy plant. Fertigation is not a new topic, but IoT-based fertigation yields better results than simple drip fertigation (Neto et al., 2014). The list of detection sensors and their applications in fertiliser application are shown in Table 3.2.

Raut et al. (2018) developed a fertigation and irrigation system using IoT. A color sensor (1185 SunRom color sensor) was employed for sensing the color of the chemical that is present in the soil sample. Nitrogen (N), phosphorus (P) and potassium (K) are major micronutrients, and sensing their status in the given soil samples saves money, time and power for the farmer. The system developed based on the status of NPK in the soil categorises the NPK requirement as high, medium and low. Rasal et al. (2017) developed an automatic soil NPK measurement and automatic fertiliser dispensing mechanism. The developed kit consists of SUNROM 1185; the NPK solution, consisting of soil pulse substances; an LPC 2148 (ARM 7) controller; an LCD display for displaying NPK values and a DC motor.

TABLE 3.2

Detection Sensors and Their Applications in Fertiliser

Sensor	Sensor Principle	Application Used
Yara N-sensor	Spectrometer	Measurement of nitrogen levels through indices
Crop Circle ACS-470	Spectrometer	Measurement of green biomass and leaf area estimates
Weedseeker	Optoelectronic sensor	Real-time spot application of agriculture chemicals
WEEDit	Optoelectronic sensor	Real-time spot application of agriculture chemicals
SUNROM 1185 color sensor	Color sensor	NPK measurement

Source: Peteinatos et al., 2014.

3.2.2 SENSORS IN DISEASE CONTROL

Insects/pests control in crops is a major activity. Timely detection of diseases/pests in a crop is key to producing a good crop yield. In the traditional method, diseases/pests are unnoticed till a spread occurs at a large scale in the field. Most of the advice on pesticide application is based on a whole area. Even if a pest appears at one spot in the field, farmers apply pesticides over the entire area. Such practices lead to the wastage of agricultural chemicals as well adverse effects on the environment. Moreover, expenditure on chemicals is also increased. The FAO estimates that 20–40% of the world's crop yields drop annually because of pests. Early detection of plant symptoms is a great challenge in the present-day agriculture field.

A study on the damage of tomato leaves by leaf miners was investigated by Xu et al. (2007) using spectral characteristics. Based on the severity level of tomato leaf damage, the infection levels were divided into five scales. A high correlation coefficient (r) was found between the sensitive bands and severity levels of 1450 and 1900 nm. Rumpf et al. (2010) used hyperspectral data for detecting powdery mildew, sugarbeet rust and leaf spot disease on sugarbeet leaves. As a classification feature, nine spectral vegetation indices were used, and the classification (infected and healthy plants) efficiency of the support vector machine was up to 97%. Moltó et al. (2001) employed ultrasonic sensors in the tree canopy volume estimation, and chemical savings were reported to be 37%. In another similar study, Solanelles et al. (2006) used ultrasonic sensors for tree canopy volume estimation of olive, pear and apple orchards and found that 28–70% spray product savings were observed compared with a conventional spraying application. Ultrasonic sensors were engaged in vineyard structure estimation, and an average of 58% chemical saving was reported by Gil et al. (2007) when compared with a uniform application rate. Thorat et al. (2017) developed an architecture for plant disease detection using IoT, and it collects real-time field information such as humidity, temperature, soil moisture and crop growth status and sends it to the farmer over Wi-Fi. The camera collects leaf images, and the DHT11 sensor collects both humidity and temperature information together. The DHT11 sensor displays the readings after 2 seconds, and its performance was found to be good when humidity and temperature ranged from 20% to 80% with an accuracy of 5% and 0°C to 50°C with an accuracy of ±2°C, respectively. All these sensor modules were connected to the Raspberry Pi. K-means clustering, an unsupervised algorithm, was employed for leaf disease prediction. An Apache server was used to obtain and send data to the website. They considered the leaf as a disease-infested leaf when the unmasked pixels were 30% more than the masked pixels. The developed system showed the best results; however, extreme sunlight decreased the image quality and affected the camera performance, and therefore classification efficiency decreased. Moreover, the system remains closed during the nighttime, which is the main limitation of the developed architecture.

3.2.3 SENSORS FOR WEED DETECTION

The spectral characteristics of vegetation are powerful parameters used for estimating biomass. The visible light wavelength ranges from 400 nm to 700 nm, whereas

TABLE 3.3
List of Commercially Available Sensors

Author Name	Sensor Name
Link et al. (2003)	WeedSeeker and GreenSeeker
Hong et al. (2007)	WEEDit
Tremblay et al. (2009)	Crop Circle ACS-470
Biller (1998)	DetectSpray

the NIR spectrum ranges from 700 nm to 2500 nm. The analysis of spectral data is a quick, non-destructive technique for the recognition and extraction of crop features. In general, green leaves show a high reflectance in the green and NIR spectra and show a low spectrum in the red and blue spectra. In the case of soil, it has high reflectance in the blue and NIR light range (Scotford and Miller, 2005; Kavdır, 2004). Identifying plants through their spectrum is relatively simple because plants are easily detectable by using their spectra. Sensors that primarily focus on very few specific spectral bands (one or two) are called optoelectronic sensors. Optoelectronic sensors principally rely on bands in the NIR and red spectrum. These sensors can discriminate plant and non-plant material by computing indices similar to the NDVI correlated with plant biomass coverage values. Such sensors cannot differentiate/discriminate between species (crops and weeds), and hence these types of sensors help detect green vegetation (weeds) present in the crop rows. Commercial sensors that are of this type measure reflectance properties in the NIR and red wavelengths to obtain an index analogous to the NDVI. Some of the optoelectronic sensors commercially available in the market are listed in Table 3.3.

Ultrasonic sensors are of low cost and have the potential to be used in various applications such as spray distance measurement (object detection and boom height and width measurement to perform uniform spraying application), canopy volume estimation (Dvorak et al., 2016) and measurement of the depth of water level in the tank. Ultrasonic sensors can be used for weed detection (Pajares et al., 2013) when combined with a camera sensor. The height of the plant is determined by using ultrasonic sensors, whereas plant and weed coverage is determined by the camera. In the case of machine vision–based weed detection (Gerhards and Oebel, 2006; Andujar et al., 2011), the employed imaging sensor, with the help of image processing tool, can not only separate green patches from soil background but also classify different weed species and, moreover, classify the main crop from weeds (Rumpf et al., 2012). Imaging sensors are also useful in plant species recognition (Zhang et al., 2008), growing phase determination (Kataoka et al., 2003) and plant disease detection (Camargo and Smith, 2009). Image acquisition, image segmentation and feature extraction (texture, shape, color) are three main parts of any image segmentation algorithm (Piron et al., 2009).

3.2.4 SENSORS IN PRECISION IRRIGATION

In conventional methods such as drip irrigation and sprinkler irrigation, irrigation is done frequently or on a scheduled basis. Due to the unplanned water use, the

TABLE 3.4
Depth of Sensor Used by Researchers in Their Studies

Sr. No.	References	Depth of Placement of Sensor
1	Kennedy et al. (2003)	Sensors at 5 cm and 15 cm respond quickly
2	Nallani and Hency (2015)	8 cm
3	Soulies et al. (2015)	Most suitable was 10 cm
4	Bowlekar et al. (2019)	Two sensors used at 5 cm and 10 cm

groundwater level is decreasing day by day. It is not an easy task to precisely estimate the water requirement of crops. Crop water requirements depend on several factors such as soil type, crop type and precipitation. Precise estimation of the water requirement of crops is possible through the deployment of sensors at the field level. The depth of the placement of the sensors used by researchers in their studies is shown in Table 3.4. Many researchers have proposed a sensor-based smart irrigation system. Different sensors are deployed in the field in a smart farming system and based on the real-time information obtained from soil moisture content, soil temperature and relative humidity (RH) sensor. A list of different types of sensors is given in Table 3.5.

3.2.5 SENSORS IN PROTECTED CULTIVATION

Greenhouse farming is considered to be the oldest method of smart farming, although the practice of cultivation of crops under a controlled environment inside the greenhouse is not new. Conversion of arable land into buildings and factories has led to seasonal weather pattern changes and food losses incurred at every stage of

TABLE 3.5
List of Sensors Engaged in Soil Moisture Estimation

Sensors Used	Conditions	Irrigation Saving (mm)	Work (year)
Soil moisture sensor (SMS)	Landscape	726	Qualls et al. (2001)
Evapotranspiration (ET) SMS	Fescue turfgrass plots	488	Grabow et al. (2008)
ET SMS	St. Augustine grass turf plots	840	McCready et al. (2009)
Self-designed wireless sensor	Virtual	685.5	Xiao et al. (2010)
Temperature sensor (RS) SMS	Bermuda grass turf plots	602	Cardenas et al. (2010)
SMS	Landscape	673	Kumar et al. (2014)
Temperature sensor (LM35) Humidity sensor (CLM53R) Soil pH sensor	Virtual	No results	Parameswaran et al. (2016)
SMS	Landscape	No results	Rawal S. (2017)

crop production, and all these factors limit the availability of a good diet. Under such a scenario, growing crops under a controlled environment significantly enhances the crop yields, and the increasing future food demand of the increasing population can be met. These greenhouse structures are used for the cultivation of fruits and vegetables during any season by controlling sunlight, inside greenhouse temperature, soil moisture and humidity.

Hydroponics, next-generation farming, is the art of soilless farming or a way of skipping the usage of soil. It involves growing a variety of crops without soil in nutrient-rich solutions in an aqueous solvent and also using a different material that supports crop roots. Concerning the IoT application in a hydroponic greenhouse, a wide variety of sensors and actuators are built in the greenhouse to collect the information from all the sensors and control environmental conditions through actuators (air pumps, irrigation pumps and other devices). Generally, in the greenhouse, seven factors are to be controlled: EC, pH, humidity, light, CO_2, temperature and dissolved oxygen. The IoT system continuously monitors the greenhouse environment based on input data of sensors. By adopting hydroponics in the greenhouse, we can save water and fertiliser use by up to 95%. It completely eliminates pesticide and herbicide use. The crucial part of this farming is the precise management of nutrients in the liquid. To take action under such a scenario, a highly reliable wireless control system has been proposed by Ibayashi et al. (2016) for tomato hydroponics. Few other control systems have also been proposed by other researchers (Nishimura et al., 2017; Theopoulos et al., 2018).

3.3 APPLICATION OF UNMANNED AERIAL VEHICLES IN PRECISION AGRICULTURE

An agricultural drone (an eye in the sky for agriculture) is a UAV used on farms to increase crop production and monitor crop growth. UAVs eliminate the problems that exist with satellites and airplanes when we consider the micro-view. The images that are obtained with UAV cameras are better than those obtained with an aircraft. UAVs are fitted with NDVI cameras that help in early and better assessment of crop health, weed coverage, soil stress and nitrogen status of crops (Cozzolino et al., 2015). In precision agriculture, drones play a significant role and combine sensor data and imaging with real-time data analytics to improve farm productivity through mapping and spatial variability in the field. When drones are coupled with thermal imaging sensors, they help recognise water quantity, as we know that plants with a high water content appear cooler in the image. The hyperspectral cameras record both visible and invisible light, identifying the specific type of plant based on the reflected light color. They also help in detecting unwanted weeds from the main crop (Adao et al., 2017). Crop-spraying drones or easy-to-fly devices are designed to spray pesticides on crops. Moreover, they can capture high-resolution images of the whole field for feature analysis. They are fit for all kinds of difficult terrains, plantations and crops of varying heights. The uses of UAVs in agriculture include crop health monitoring, agriculture chemical spraying, planting and plant counting and aerial photography.

3.4 APPLICATION OF SMARTPHONES IN PRECISION AGRICULTURE

The smartphone is now a widespread communications device that is available on every corner of the globe and available at the rural level as well. Smartphones consist of different sensors such as a built-in cameras, GPS sensors, proximity sensors and temperature sensors. The list of sensors present in smartphone devices is given in Figure 3.3. Smartphones have high data processing and storage capabilities. Therefore, they can play an important role in real-time information gathering on weather, crop conditions and soil health information. Smartphone has the potential to be used as an IoT tool, especially at the rural level. Applications of the mobile camera sensor in agriculture are mainly soil color determination (Robledo et al., 2013), determination of carbon content and organic matter content in the topsoil (Aitkenhead et al., 2013), prediction of chlorophyll content of the rice leaves (Intaravanne and Sumriddetchkajorn, 2012), measurement of ripening status of fruit under UV-A light source and white light source (Sumriddetchkajorn, 2013), leaf area index determination using images of leaf canopy and fruit size estimation (Wang et al., 2018).

Sumriddetchkajorn (2013) developed the mobile application BaiKhaoNK to evaluate the color level of rice leaves and recommend the required amounts of nitrogen fertiliser to be applied to the rice field. Pérez-Castro et al. (2017) developed a mobile app called cFerigUAL for fertigation. This app recommends the amount of water and fertiliser to be applied based on various fertigation technologies and various crop growing systems. Liu and Koc (2013) developed SafeDriving, an iOS mobile application that calculates the stability index using the accelerometer, gyroscope, and GPS to detect tractor rollover. While tractor-traveling, if any changes in the stability of tractor are noticed, then the mobile application immediately alerts the driver. If the stability index crosses the threshold value, there is a warning alarm

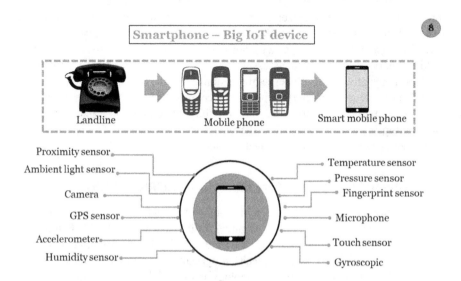

FIGURE 3.3 List of sensors present in the recently launched new smartphone.

TABLE 3.6

Application of Smartphones in Precision Agriculture

Author	Sensor	Application and Feature
Prasad et al. (2014)	Camera	Disease diagnosis: Leaf images acquired using mobile camera and acquired images sent to the laboratory for further disease diagnosis.
Gómez-Robledo et al. (2013)	Camera	Soil color sensor: A mobile application developed for soil color estimation. Once soil image is captured, the user needs to select interested area in the image and the respective color name of the interested area is displayed on the screen.
Ferguson et al. (2016)	Camera	A spraying application called SnapCard was developed based on image processing analysis for a field in-field analysis of spray collectors. It follows five imaging techniques to quantify droplet size and deposition.
Aitkenhead et al. (2013)	GPS	Information on soil parameters such as soil carbon, pH and NPK are delivered based on user's GPS location.
Wu and Chang (2013)	GPS	Pest or disease information is delivered based on GPS location.
Saha et al. (2012)/Jagyasi et al. (2011)	Microphone	m-Sahayak/mKRISHI are mobile applications designed for farmers, who use the microphone for recording their queries and send them to experts.
Liu and Koc (2013)	Accelerometer/ gyroscope	SafeDriving: A safe driving mobile application. The roll and pitch angles of a tractor are obtained in order to estimate the rollover event of a tractor.

pop-up in the mobile phone. If the tractor is subjected to accident or toppling, then the stability index becomes zero. The developed mobile application immediately sends information about the accident location and time to the emergency contacts list. Wang et al. (2018) developed a mobile application for fruit health assessment and yield estimation with a typical assessment rate of 240 fruits per hour. OpenCV was used for image processing, object segmentation and fruit size estimation. The outside illumination intensity affected the performance of fruit estimation using the camera. Another limitation of the study was holding the mobile phone at a fixed height and maintaining it perfectly vertical to the object to be measured. The application of smartphones in various fields is shown in Table 3.6.

3.5   APPLICATION OF IoT IN PRECISION AGRICULTURE

The IoT, artificial intelligence (AI) and big data have recently made extraordinary progress in many industries, including agriculture. Still, when we think of IoT in agriculture, communication facilities such as base stations or Wi-Fi are very limited. Wireless communication protocols used in precision agriculture are given in Table 3.7. This unavailability of communication technology is one of the major drawbacks while implementing IoT in the agriculture field, which prevents the IoT's growth in this

TABLE 3.7

Wireless Communication Protocols Used in Precision Agriculture (PA)

	ZigBee	Bluetooth	Bluetooth Wibree	Wi-Fi
Frequency band	2.4 GHz range	2.4 GHz	2.4 GHz	2.4 GHz
Range	30 m–1.6 km	30–300 ft	Up to 10 ft	100–150 ft
Data rate	250 kbps	1 Mbps	1 Mbps	11–54 Mbps
Power consumption	Low	Medium	Low	High
Cost	Low	Low	Low	Low
Modulation/protocol	DSSS, CSMA/CA	FHSS	FHSS	DSSS/CCK, OFDM
Security	128 bit	64 or 128 bit	128 bit	128 bit

Source: Rehman et al., 2008.

sector. The drawback of the IoT in Indian agriculture is that most farmers are not aware of these sensors' roles, the IoT and their application in precision agriculture (Abbasi et al., 2014). Other drawbacks include the lack of IT infrastructure facilities, electric power and Internet supply to remote places and the landholding capacity of Indian farmers being very small compared to developed countries. Many companies offer IoT solutions to the farmers, and a list of a few companies is given in Table 3.8.

3.6 CONCLUSION

The application of sensors in precision agriculture brings an increase in revenue generation and almost completely eliminates human involvement, thereby leading to a reduction in stress for the farmer and a decrease in energy use. The smartphone,

TABLE 3.8

Precision Agriculture—IoT Solutions from Different Companies

Company	Product	Source
John Deer	Self-driving tractor, combine harvester, smart sprayer	https://digital.hbs.edu/platform-digit/submission/farm-to-data-table-john-deere-and-data-in-precision-agriculture/
Case IH	Autonomous self-driving tractor	https://www.caseih.com/northamerica/en-us/Pages/campaigns/autonomous-concept-vehicle.aspx
The French drone maker, Parrot	Drones and multispectral sensors	https://www.parrot.com/us
Connecterra	Dairy cattle tracking monitoring, monitoring animal health and birthing timings	https://www.connecterra.io/
CropX	Wireless sensors for monitoring soil moisture	https://www.cropx.com/
Deutsche Telekom (DT)	Real-time operation of agricultural machinery	https://www.telekom.com/en

a low-cost and handheld device, consists of a variety of sensors, and they also play a vital role in precision agriculture. A benefit of precision agriculture is that the advanced sensor technologies can be easily incorporated into existing machines without major modifications. Precision agriculture and conservation agriculture are possible only through the deployment of this advanced sensor technologies in present-day agriculture. Today, various sensors and actuators are available in the market that can be easily deployed in the agriculture field for variable rate application of pesticide or fertilisers, greenhouse automation and monitoring and yield monitoring. Variable rate technology eliminates the excess application of pesticides or fertilisers and saves the environment from pollution. In the future, IoT is going to play a major role in precision agriculture with the upcoming 5G technology. An IoT-based precision irrigation system can give an optimal solution for autonomous agricultural operations for precision, economics, reduced human struggle and environment protection.

REFERENCES

Abbasi, A. Z., Islam, N., & Shaikh, Z. A. (2014). A review of wireless sensors and networks' applications in agriculture. *Computer Standards & Interfaces*, 36(2), 263–270.

Adão, T., Hruška, J., Pádua, L., Bessa, J., Peres, E., Morais, R., & Sousa, J. J. (2017). Hyperspectral imaging: A review on UAV-based sensors, data processing and applications for agriculture and forestry. *Remote Sensing*, 9(11), 1110.

Aitkenhead, M., Donnelly, D., Coull, M., & Black, H. (2013, October). E-smart: Environmental sensing for monitoring and advising in real-time. In *International Symposium on Environmental Software Systems* (pp. 129–142). Springer, Berlin, Heidelberg.

Andujar, D., Angela Ribeiro., Fernandez-Quintanilla, C., & Dorado J. (2011). Accuracy and feasibility of optoelectronic sensors for weed mapping in wide row crops. *Sensors*, 11, 2304–2318.

Anonymous 2020a. Department of chemical and petro chemical, ministry of chemical and fertilizer, Government of India (GOI), as on 18.07.2019. https://chemicals.nic.in/.

Biller, R. H. (1998). Reduced input of herbicides by use of optoelectronic sensors. *Journal of Agricultural Engineering Research*, 71(4), 357–362.

Bowlekar, A. P., Patil, S. T., Kadam, U. S., Mane, M. S., Nandgude, S. B., & Palte, N. K. (2019). Development, field testing and economic evaluation of automatic irrigation system. *Journal of Agricultural Engineering*, 56(4), 284–293.

Brown, D. L., Giles, D.K., Oliver, M.N., & Klassen, P. (2008). Targeted spray technology to reduce pesticide in runoff from dormant orchards. *Crop Protection*; 27(3–5), 545–552.

Camargo, A., & Smith, J. S. (2009). An image-processing based algorithm to automatically identify plant disease visual symptoms. *Biosystems Engineering*, 102(1), 9–21.

Cardenas-Lailhacar, B., Dukes, M. D., & Miller, G. L. (2010). Sensor-based automation of irrigation on bermudagrass during dry weather conditions. *Journal of Irrigation and Drainage Engineering*, 136(3), 184–193.

Cozzolino, D., Porker, K., & Laws, M. (2015). An overview on the use of infrared sensors for in field, proximal and at harvest monitoring of cereal crops. *Agriculture*, 5(3), 713–722.

Dvorak, J. S., Stone, M. L., & Self, K. P. (2016). Object detection for agricultural and construction environments using an ultrasonic sensor. *Journal of Agricultural Safety and Health*, 22(2), 107–119.

Ferguson, J. C., Chechetto, R. G., O'Donnell, C. C., Fritz, B. K., Hoffmann, W. C., Coleman, C. E., ... & Hewitt, A. J. (2016). Assessing a novel smartphone application–SnapCard, compared to five imaging systems to quantify droplet deposition on artificial collectors. *Computers and Electronics in Agriculture*, 128, 193–198.

Gerhards, R., & Oebel, H. (2006). Practical experiences with a system for site-specific weed control in arable crops using real-time image analysis and GPS-controlled patch spraying. *Weed Research*, 46(3), 185–193.

Gil, E., Escolà, A., Rosell, J. R., Planas, S., & Val, L. (2007). Variable rate application of plant protection products in vineyard using ultrasonic sensors. *Crop Protection*, 26(8), 1287–1297.

Gómez-Robledo, L., López-Ruiz, N., Melgosa, M., Palma, A. J., Capitán-Vallvey, L. F., & Sánchez-Marañón, M. (2013). Using the mobile phone as Munsell soil-colour sensor: An experiment under controlled illumination conditions. *Computers and Electronics in Agriculture*, 99, 200–208.

Grabow, G. L., Vasanth, A., Bowman, D., Huffman, R. L., & Miller, G. L. (2008). Evaluation of evapotranspiration-based and soil-moisture-based irrigation control in turf. In *World Environmental and Water Resources Congress 2008: Ahupua'A* (pp. 1–9).

Hong, S. D., Schepers, J. S., Francis, D. D., & Schlemmer, M. R. (2007). Comparison of ground-based remote sensors for evaluation of corn biomass affected by nitrogen stress. *Communications in Soil Science and Plant Analysis*, 38(15-16), 2209–2226.

Ibayashi, H., Kaneda, Y., Imahara, J., Oishi, N., Kuroda, M., & Mineno, H. (2016). A reliable wireless control system for tomato hydroponics. *Sensors*, 16(5), 644.

Intaravanne, Y., & Sumriddetchkajorn, S. (2012, November). BaiKhao (rice leaf) app: A mobile device-based application in analyzing the color level of the rice leaf for nitrogen estimation. In *Optoelectronic Imaging and Multimedia Technology II* (Vol. 8558, p. 85580F). International Society for Optics and Photonics.

Jagyasi, B. G., Pande, A. K., & Jain, R. (2011, October). Event based experiential computing in agro-advisory system for rural farmers. In *2011 IEEE 7th International Conference on Wireless and Mobile Computing, Networking and Communications (WiMob)* (pp. 439–444). IEEE.

Kataoka, T., Kaneko, T., Okamoto, H., & Hata, S. (2003, July). Crop growth estimation system using machine vision. In *Proceedings 2003 IEEE/ASME International Conference on Advanced Intelligent Mechatronics (AIM 2003)* (Vol. 2, pp. b1079–b1083). IEEE.

Kavdır, İ. (2004). Discrimination of sunflower, weed and soil by artificial neural networks. *Computers and Electronics in Agriculture*, 44(2), 153–160.

Kennedy, J. R., Keefer, T. O., Paige, G. B., & Barnes, E. (2003). Evaluation of dielectric constant-based soil moisture sensors in a semiarid rangeland. In *Proceedings First Interagency Conference on Research in the Watersheds* Washington, DC: Consortium of Universities for the Advancement of Hydrologic Science, Inc (pp. 503–508). October 27-30, 2003. U.S. Department of Agriculture, Agricultural Research Service.

Kumar, A., Kamal, K., Arshad, M. O., Mathavan, S., & Vadamala, T. (2014, October). Smart irrigation using low-cost moisture sensors and XBee-based communication. In *IEEE Global Humanitarian Technology Conference (GHTC 2014)* (pp. 333–337). IEEE.

Link, A., Panitzki, M., & Reusch, S. (2003). Hydro N-Sensor: tractor-mounted remote sensing for variable nitrogen fertilization. In *Proceedings of the 6th International Conference on Precision Agriculture and Other Precision Resources Management, Minneapolis, MN, USA, 14-17 July, 2002* (pp. 1012-1017). American Society of Agronomy.

Liu, B., & Koc, A. B. (2013). SafeDriving: A mobile application for tractor rollover detection and emergency reporting. *Computers and Electronics in Agriculture*, 98, 117–120.

Liu, H., Wang, X., & Bing-kun, J. (2018). Study on NDVI optimization of corn variable fertilizer applicator. *INMATEH-Agricultural Engineering*, 56(3), 193–202.

McCready, M. S., Dukes, M. D., & Miller, G. L. (2009). Water conservation potential of smart irrigation controllers on St. Augustinegrass. *Agricultural Water Management*, 96(11), 1623–1632.

Moltó, E., MartíN, B., & Gutiérrez, A. (2001). Pm—Power and machinery: Pesticide loss reduction by automatic adaptation of spraying on globular trees. *Journal of Agricultural Engineering Research*, 78(1), 35–41.

Nallani, S., & Hency, V. B. (2015). Low power cost effective automatic irrigation system. *Indian Journal of Science and Technology, 8*(23), 1.

Navulur, S., & Prasad, M. G. (2017). Agricultural management through wireless sensors and Internet of Things. *International Journal of Electrical and Computer Engineering, 7*(6), 3492.

Neto, A. J. S., Zolnier, S., & de Carvalho Lopes, D. (2014). Development and evaluation of an automated system for fertigation control in soilless tomato production. *Computers and Electronics in Agriculture, 103*, 17–25.

Nishimura, T., Okuyama, Y., Matsushita, A., Ikeda, H., & Satoh, A. (2017, October). A compact hardware design of a sensor module for hydroponics. In *2017 IEEE 6th Global Conference on Consumer Electronics (GCCE)* (pp. 1–4). IEEE.

Pajares, G., Peruzzi, A., & Gonzalez-de-Santos, P. (2013). Sensors in agriculture and forestry. *Sensors. 13*(9):12132–12139.

Parameswaran, G., & Sivaprasath, K. (2016). Arduino based smart drip irrigation system using Internet of Things. *International Journal of Engineering Science and Computing, 6*, 5518.

Pérez-Castro, A., Sánchez-Molina, J. A., Castilla, M., Sánchez-Moreno, J., Moreno-Úbeda, J. C., & Magán, J. J. (2017). cFertigUAL: A fertigation management app for greenhouse vegetable crops. *Agricultural Water Management, 183*, 186–193.

Peteinatos, G. G., Weis, M., Andújar, D., Rueda Ayala, V., & Gerhards, R. (2014). Potential use of ground-based sensor technologies for weed detection. *Pest Management Science, 70*(2), 190–199.

Piron, A., Leemans, V., Lebeau, F., & Destain, M. F. (2009). Improving in-row weed detection in multispectral stereoscopic images. *Computers and Electronics in Agriculture, 69*(1), 73–79.

Prasad, S., Peddoju, S. K., & Ghosh, D. (2014, April). Energy efficient mobile vision system for plant leaf disease identification. In *2014 IEEE Wireless Communications and Networking Conference (WCNC)* (pp. 3314–3319). IEEE.

Qualls, R. J., Scott, J. M., & DeOreo, W. B. (2001). Soil moisture sensors for urban landscape irrigation: Effectiveness and reliability 1. *JAWRA Journal of the American Water Resources Association, 37*(3), 547–559.

Rasal, P. M., Tilekar, S. B., Todkar, A. D., & Jagtap, S. A. (2017). Nek soil measurement and automatic fertilizer dispense. *International Journal for Research in Applied Science and Engineering Technology, 5*(3), 806–809.

Raut, R., Varma, H., Mulla, C., & Pawar, V. R. (2018). Soil monitoring, fertigation, and irrigation system using IoT for agricultural application. In *Intelligent Communication and Computational Technologies* (pp. 67–73). Springer, Singapore.

Rawal, S. (2017). IOT based smart irrigation system. *International Journal of Computer Applications, 159*(8), 7–11.

Renard, Kenneth G., McElroy, Stephen A., Gburek, William J., Canfield, H. Evan and Scott, Russell L., eds. 2003. First Interagency Conference on Research in the Watersheds, October 27-30, 2003. U.S. Department of Agriculture, Agricultural Research Service.

Rumpf, T., Mahlein, A. K., Steiner, U., Oerke, E. C., Dehne, H. W., & Plümer, L. (2010). Early detection and classification of plant diseases with support vector machines based on hyperspectral reflectance. *Computers and Electronics in Agriculture, 74*(1), 91–99.

Rumpf, T., Römer, C., Weis, M., Sökefeld, M., Gerhards, R., & Plümer, L. (2012). Sequential support vector machine classification for small-grain weed species discrimination with special regard to Cirsium arvense and Galium aparine. *Computers and Electronics in Agriculture, 80*, 89–96.

Saha, B., Ali, K., Basak, P., & Chaudhuri, A. (2012). Development of m-Sahayak-the innovative Android based application for real-time assistance in Indian agriculture and health sectors. In *The Sixth International Conference on Mobile Ubiquitous Computing, Systems, Services and Technologies* (pp. 133–137). September 23–28, Barcelona, Spain.

Scotford, I. M., & Miller, P. C. H. (2005). Applications of spectral reflectance techniques in northern European cereal production: A review. *Biosystems Engineering, 90*(3), 235–250.

Solanelles, F., Escolà, A., Planas, S., Rosell, J. R., Camp, F., & Gràcia, F. (2006). An electronic control system for pesticide application proportional to the canopy width of tree crops. *Biosystems Engineering, 95*(4), 473–481.

Soulis, K. X., Elmaloglou, S., & Dercas, N. (2015). Investigating the effects of soil moisture sensors positioning and accuracy on soil moisture based drip irrigation scheduling systems. *Agricultural Water Management, 148*, 258–268.

Sumriddetchkajorn, S. (2013, June). How optics and photonics is simply applied in agriculture? In *ICPS 2013: International Conference on Photonics Solutions* (Vol. 8883, p. 888311). International Society for Optics and Photonics.

Theopoulos, A., Boursianis, A., Koukounaras, A., & Samaras, T. (2018, May). Prototype wireless sensor network for real-time measurements in hydroponics cultivation. In *2018 7th International Conference on Modern Circuits and Systems Technologies (MOCAST)* (pp. 1–4). IEEE.

Thorat, A., Kumari, S., & Valakunde, N. D. (2017, December). An IoT based smart solution for leaf disease detection. In *2017 International Conference on Big Data, IoT and Data Science (BID)* (pp. 193–198). IEEE.

Tremblay, N., Wang, Z., Ma, B. L., Belec, C., & Vigneault, P. (2009). A comparison of crop data measured by two commercial sensors for variable-rate nitrogen application. *Precision Agriculture, 10*(2), 145–161.

Wang, Z., Koirala, A., Walsh, K., Anderson, N., & Verma, B. (2018). In field fruit sizing using a smart phone application. *Sensors, 18*(10), 3331.

Wu, Y., & Chang, K. T. (2013). An empirical study of designing simplicity for mobile application interaction. *19th Americas Conference on Information Systems, AMCIS 2013 - Hyperconnected World: Anything, Anywhere, Anytime* 1: 331–338. ScholarBank@ NUS Repository.

Xiao, K., Xiao, D., & Luo, X. (2010). Smart water-saving irrigation system in precision agriculture based on wireless sensor network. *Transactions of the Chinese Society of Agricultural Engineering, 26*(11), 170–175.

Xu, H. R., Ying, Y. B., Fu, X. P., & Zhu, S. P. (2007). Near-infrared spectroscopy in detecting leaf miner damage on tomato leaf. *Biosystems Engineering, 96*(4), 447–454.

Zhang, L., Kong, J., Zeng, X., & Ren, J. (2008, October). Plant species identification based on neural network. In *2008 Fourth International Conference on Natural Computation* (Vol. 5, pp. 90–94). IEEE.

4 Perspectives on Deep Learning Techniques for Industrial IoT

R. Dhaya, R. Kanthavel and Harun Bangali

CONTENTS

4.1 INTRODUCTION

Firms are progressively handling information gathered by Internet of Things (IoT) sensors into machine learning (ML) models and utilising the subsequent data to change how they work and the items and administrations they offer. As amounts of Web-related sensors are consolidated into vehicles, planes preparation and structures, associations are saving enormous proportions of data. Exploiting this data to isolate accommodating information is a test that is starting to be met using the model planning limits of ML—a subset of the field of artificial intelligence (AI). Firms are progressively taking care of information gathered by IoT sensors—arranged wherever from ranchers' fields to train tracks—into ML models and utilising the subsequent data to improve their business cycles, items and administrations (M. S. Mahdavinejad et al. 2018).

The adoption of inventive-advanced innovations likewise alluded to as industry is dynamically improving item quality, work security, correcting flawed forecasts and increasing productivity in energy use and creation. Industryies' ideas are relied upon to build their impression in modern areas by 20% in the future since they permit concise and more proficient creation. In this unique situation, many assembling organisations are keen on quickening the selection and reconciliation of secure, dependable AI. Specifically, AI-based assembling can improve business key execution pointers of gathering measures by utilising heterogeneous modern huge information investigation, data displaying and alliance. In the specific circumstance, the interconnection of AI-based assembling measures with currently conveyed remote organisations is a difficult examination field, particularly when focal preparing is performed outside mechanical premises. Nevertheless, most AI methods depend on numerical models that are hard to comprehend by the overall population, so the vast majority use AI-based innovation as a black box that they ultimately begin to confide independent of their insight. The application of Human-Driven AI (H-AI) in IoT frameworks is another area of investigation, as IoT frameworks cannot only benefit from customers but must also provide straightforward clarifications about choices or evaluations (Y. Xu et al. 2018).

The industrial IoT (IIoT) is an actual organisation of things, items, or gadgets (that contain installed innovation) for detecting and controlling, in a mechanical setting that permits a more prominent mix among physical and digital universes. In the Fifth-Age (5G), time, proficient, dependable and superior applications' arrangement must be joined with the abuse of capacities offered by 5G organisations. Ideal utilisation of the available assets must be acknowledged while ensuring exacting Quality of Service (QoS) necessities, for example, high information rates, super low dormancy and jitter. In this specific circumstance, AI is basic to the online protection part of an IIoT-empowered associated climate, for precisely distinguishing and relieving dangers (L. Da Xu et al. 2014). Simultaneously, acquainting AI will lead to a more profitable and secure working space, calming human laborers from routine methodology and utilising canny machines and robots to perform substantial errands, subsequently permitting human specialists to zero in imagination, thinking and dynamic (Y. LeCun et al. 2015).

Notwithstanding, the reception and incorporation of AI-based advancement in assembling domains accompany a couple of obstacles and admonitions that should be appropriately tended to exploit its maximum capacity, without endangering indispensable jobs of people and the assurance of delicate information and methods. In this unique circumstance, the objective of the examination introduced in this original copy is to dissect an AI-based coordinated effort approach in mechanical IoT production. On the other hand, by empowering simple admittance to, and association with a wide assortment of actual gadgets or things, for example, vehicles, machines, clinical sensors, IoT encourages the improvement of utilisation in a wide range of domains. The accompanying graph features the key application domains of IoT.

Medical services, Industry 4.0, energy executives and shrewd matrices, transportation, savvy foundation, retail, and a variety of other areas are among those that will improve our lives and social orders. These applications will have a worldwide financial effect of $4–$11 trillion every year by 2025. The key givers of this amount of cash incorporate the accompanying of plants or ventures, including activities of the board and prescient maintenance; urban areas, including public security, wellbeing, traffic signal and asset the executives; medical care, including checking and overseeing diseases and improving health; retail, including self-checkouts and stock administration; and energy, including a brilliant network. Table 4.1 describes the summary of literature surveys.

TABLE 4.1
Summary of Literature Surveys

Authors	Topic	Method/Inference
K. Alrawashdeh and C. Purdy, 2016	Toward an Online Anomaly Intrusion Detection System Based on Deep Learning	A deep learning approach for anomaly detection using a restricted Boltzmann machine (RBM) and a deep belief network
B.-H. Li, B.-C. Hou, W.-T. Yu, X.-B. Lu and C.-W. Yang, 2017	Applications of Artificial Intelligence in Intelligent Manufacturing	Integration of AI technology with information communications, manufacturing
B. Chen and J. Wan, 2019	Emerging Trends of ML-Based Intelligent Services for the Industrial Internet of Things	ML-based IIoT architecture for intelligent IIoT services
A. Thantharate, R. Paropkari, V. Walunj and C. Beard, 2019	DeepSlice: A Deep Learning Approach towards an Efficient and Reliable Network Slicing in 5G Networks	Machine learning (ML) and deep learning (DL) neural network models
P. Zheng, Z. Sang, R. Y. Zhong, Y. Liu, C. Liu, K. Mubarok, S. Yu and X. Xu, 2018	Smart Manufacturing Systems for Industry 4.0: Conceptual Framework, Scenarios and Future Perspectives	A conceptual framework of smart manufacturing systems for Industry 4.0
F. Ullah, H. Naeem, S. Jabbar, S. Khalid, M. A. Latif, F. Al-Tudjman and L. Mostarda, 2019	Cyber Security Threats Detection in the Internet of Things Using Deep Learning Approach	Deep convolution neural network
Q. Zhang, L. T. Yang, Z. Chen, P. Li and F. Bu, 2018	An Adaptive Dropout Deep Computation Model for Industrial IoT Big Data Learning with Crowdsourcing to Cloud Computing	Deep computation model—outsourcing selection algorithm based on the maximum entropy
Y. Xu, S. Li, D. Zhang, Y. Jin, F. Zhang, N. Li and H. Li, 2018	Identification Framework for Cracks on a Steel Structure Surface by a Restricted Boltzmann Machine Algorithm Based on Consumer-Grade Camera Images	Framework based on a restricted Boltzmann machine (RBM)
Y. Han, C.-J. Zhang, L. Wang and Y.-C. Zhang, 2019	Industrial IoT for Intelligent Steelmaking with Converter Mouth Flame Spectrum Information Processed by Deep Learning	Dynamic prediction model—carbon content and temperature value
K. Lepenioti, M. Pertselakis, A. Bousdekis, A. Louca, F. Lampathaki, D. Apostolou, G. Mentzas and S. Anastasiou, 2020	Machine Learning for Predictive and Prescriptive Analytics of Operational Data in Smart Manufacturing	Recurrent neural networks for predictive analytics and multi-objective reinforcement learning

(Continued)

TABLE 4.1 *(Continued)*
Summary of Literature Surveys

Authors	Topic	Method/Inference
F. Liang, W. Yu, X. Liu, D. Griffith and N. Golmie, 2020	Toward Edge-Based Deep Learning in Industrial Internet of Things	Convolutional neural network (CNN) model, utilising a real-world IIoT data set to evaluate
P. Lade, R. Ghosh and S. Srinivasan, 2017	Manufacturing Analytics and Industrial Internet of Things	Case study and provides detail about challenges and approaches in data extraction, modeling and visualisation
S. Jeschke, C. Brecher, T. Meisen, D. Ozdemir and T. Eschert, 2017	Industrial Internet of Things and Cyber-Manufacturing Systems	Future trends of IIoT and CMS within Industry 4.0
S. Prathibha, A. Hong and M. Jyothi, 2017	IoT-Based Monitoring System in Smart Agriculture	IoT and smart agriculture using automation
S. Sudha, M. Vidhyalakshmi, K. Pavithra, K. Sangeetha and V. Swathi, 2016	An Automatic Classification Method for Environment: Friendly Waste Segregation Using Deep Learning	Automated recognition system using deep learning algorithm in artificial intelligence to classify objects as biodegradable and nonbiodegradable
H. Yan, J. Wan, C. Zhang, S. Tang, Q. Hua and Z. Wang, 2018	Industrial Big Data Analytics for Prediction of Remaining Useful Life Based on Deep Learning	Device electrocardiogram (DECG) and an algorithm based on deep denoising autoencoder (DDA)
J. Wang, Y. Ma, L. Zhang, R. X. Gao and D. Wu, 2018	Deep Learning for Smart Manufacturing: Methods and Applications	A comprehensive survey of commonly used deep learning algorithms
R. A. Khalil, E. Jones, M. I. Babar, T. Jan, M. H. Zafar and T. Al-Hussain, 2019	Speech Emotion Recognition Using Deep Learning Techniques: A Review	Speech emotion recognition (SER)
P. Li, Z. Chen, L. T. Yang, Q. Zhang and M. J. Deen, 2017	Deep Convolutional Computation Model for Feature Learning on Big Data in the Internet of Things	Deep convolutional computation model (DCCM)
M. K. Putchala, 2017	Deep Learning Approach for Intrusion Detection System (IDS) in the Internet of Things (IoT) Network Using Gated Recurrent Neural Networks (GRNs)	A light-weight and multilayered design for an IoT network
C. Qiu, F. R. Yu, H. Yao, C. Jiang, F. Xu and C. Zhao, 2018	Blockchain-Based Software-Defined Industrial Internet of Things: A Dueling Deep Q-Learning Approach	Blockchain-based consensus protocol in SDIIoT
A. Essien and C. Giannetti, 2020	A Deep Learning Model for Smart Manufacturing Using Convolutional LSTM Neural Network Autoencoders	Deep convolutional LSTM encoder-decoder
B. Roy and H. Cheung, 2018	A Deep Learning Approach for Intrusion Detection in the Internet of Things Using Bi-Directional Long Short-Term Memory Recurrent Neural Network	Novel deep learning technique for detecting attacks within the IoT network using bidirectional long short-term memory recurrent neural network (BLSTM RNN)

(Continued)

TABLE 4.1 *(Continued)*
Summary of Literature Surveys

Authors	Topic	Method/Inference
X.-B. Jin, N.-X. Yang, X.-Y. Wang, Y.-T. Bai, T.-L. Su and J.-L. Kong, 2020	Hybrid Deep Learning Predictor for Smart Agriculture Sensing Based on Empirical Mode Decomposition and Gated Recurrent Unit Group Model	Empirical mode decomposition (EMD) method
T. Wong and Z. Luo, 2018	Recurrent Auto-Encoder Model for Large-Scale Industrial Sensor Signal Analysis	Using rolling fixed window approach
L. Ren, Y. Sun, J. Cui and L. Zhang, 2018	Bearing Remaining Useful Life Prediction Based on Deep Autoencoder and Deep Neural Networks	A novel eigenvector based on time-frequency-wavelet joint features
L. Wen, L. Gao and X. Li, 2017	A New Deep Transfer Learning Based on Sparse Auto encoder for Fault Diagnosis	New deep transfer learning
Y. Meidan, M. Bohadana, Y. Mathov, Y. Mirsky, A. Shabtai, D. Breitenbacher and Y. Elovici, 2018	N-Bait? Network-Based Detection of IoT Botnet Attacks Using Deep Autoencoders	A novel network-based anomaly detection method for the IoT called N-BaIoT IoT devices
M. Liu, F. R. Yu, Y. Teng, V. C. Leung and M. Song, 2019	Performance Optimisation for Blockchain-Enabled Industrial Internet of Things (IIoT) Systems: A Deep Reinforcement Learning Approach	Blockchain is considered as a promising solution to enable data storing/processing/sharing in a secure and efficient way
L. Zeng, E. Li, Z. Zhou and X. Chen, 2019	Boomerang: On-Demand Cooperative Deep Neural Network Inference for Edge Intelligence on the Industrial Internet of Things	Boomerang exploits DNN right-sizing and DNN partition to execute DNN inference tasks with low latency as well as high accuracy
A. Hassanzadeh, S. Modi and S. Mulchandani, 2015	Towards Effective Security Control Assignment in the Industrial Internet of Things	The new framework for analysing the effectiveness of security controls in IIoT environment
M. Ranzato, Y.-L. Boureau and Y. L. Cun, 2008	Sparse Feature Learning for Deep Belief Networks	Extracting features from a dataset of handwritten numerals, and a dataset of natural image patches
A.-H. Muna, N. Moustafa and E. Sitnikova, 2018	Identification of Malicious Activities in Industrial Internet of Things Based on Deep Learning Models	An anomaly detection technique for IICSs based on deep learning models
Z. Liang, G. Zhang, J. X. Huang and Q. V. Hu, 2014	Deep Learning for Healthcare Decision Making with EMRs	The modified version of convolutional deep learning
N. Mehdiyev, J. Lahann, A. Emrich, D. Enke, P. Fettke and P. Loos, 2017	Time Series Classification Using Deep Learning for Process Planning: A Case from the Process Industry	Framework on sensor time-series data from the process industry
F. P. Carvalho, 2017	Mining Industry and Sustainable Development: Time for Change	Mining legacy and environmental remediation
Z. Li, J. Kang, R. Yu, D. Ye, Q. Deng and Y. Zhang, 2017	Consortium Blockchain for Secure Energy Trading in Industrial Internet of Things	Optimal pricing strategy using Stackelberg game for credit-based loans

4.2 DATA ANALYSIS AND VISUALISATION FOR MACHINE LEARNING PROCESS

Data visualisation is the main part of information science. It is one of the fundamental instruments used to break down and study connections between various factors. Information representation can be utilised for engaging examination (Y. Han et al. 2019). Information perception is additionally utilised in AI for information preprocessing and examination; highlight choice; model structure; model testing; and model assessment. Placing information into the setting implies that you will envision all the sections inside the information to comprehend the accompanying: Figure 4.1 shows the data visualisation process (M. M. Najafabadi et al. 2015).

- The importance of every section of information.
- Whether it's a downright or constant variable for every section.
- Whether the information is an autonomous or ward variable.

Information Analysis: We use information representation to consider connections between highlights, just as the relationship between indicator factors and the objective variable.

Model Building: A straightforward direct relapse model could then be worked out to foresee the group variables utilising the four objective factors: tonnage, travelers, length and lodges. Once more, we can utilise information representation to analyse the real and anticipated team esteem (F. Zantalis et al. 2019).

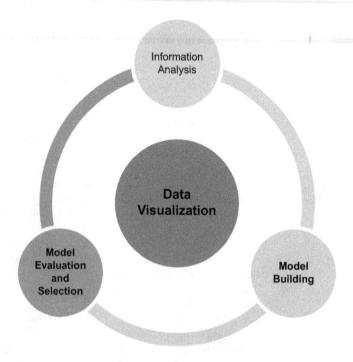

FIGURE 4.1 Data visualisation process.

Model Evaluation and Selection: We can execute three diverse relapse models, including linear regression (LR), k-nearest neighbors regression (kNR) and support vector regression (SVR); and afterward utilising the cross-approval technique, one could compute the mean cross-approval score as it's a measurement for evaluating vulnerabilities in the normal R2 score esteem.

Information perception assumes a fundamental job in the AI cycle from information preprocessing, to information investigation, to highlight choice, to demonstrate working, to show testing and model assessment (K. Lepenioti et al. 2020). It is suggested that information representation be utilised all through the AI cycle when fabricating a model, as it encourages one to imagine halfway outcomes to guarantee that there are no mistakes or irregularities in the model structure measure. Table 4.2 elaborates the Machine Vision Use Cases, Industrial Revolutions, Enabling Technologies, Practical Challenges, Main Activities, Machine Vision Applications and Machine Vision System goals.

4.2.1 How is Machine Vision Used in Manufacturing?

Machine vision applications are most proficient when deliberately coordinated into the shrewd plant as it upgrades both human and computerised execution. Makers should zero in on these two perspectives to use machine vision advantages without limit: How the machine vision innovation can empower administrators to be more proficient and precise and how different advances can consistently incorporate with machine vision to open the maximum capacity. Here the table shows the four machine vision use cases for the assembling business.

4.3 INDUSTRIAL IoT

The IIoT insinuates the augmentation and utilisation of the IoT in mechanical regions and applications (E. Sisinni et al. 2018). With a strong focus on machine-to-machine (M2M) correspondence, colossal information and AI, the IIoT enables adventures and endeavors to have better adequacy and steady quality in their undertakings. The IIoT joins modern applications, including mechanical innovation, clinical devices and programming-portrayed creation measures (J. Gubbi et al. 2013). Unsteady IIoT structures can incite operational unsettling influence and monetary hardship, among other broad outcomes. More related conditions mean more noteworthy security perils, for instance, Figure 4.2 shows the security risks in IIoT (F. Liang et al. 2020).

- Programming shortcomings that can be mishandled to attack structures
- Publicly available Web-related devices and structures
- Malicious practices like hacking, coordinated attacks and information breaks
- System control that can cause operational interruption or mischief measures
- System glitch that can achieve mischief of contraptions and genuine workplaces or injury to heads or people close by
- OT systems held for coercion, as sabotaged through the IT atmosphere

TABLE 4.2
Machine Vision Use Cases, Industrial Revolutions, Enabling Technologies, Practical Challenges, Main Activities, Machine Vision Applications and Machine Vision System Goals

Machine Vision Use Cases for the Manufacturing Industry

Use Case	Purpose
Taking proactive measures	Manufacturing companies deal with large-scale production of goods, several large machinery or heavy-duty equipment has to be used
Inspecting goods	Or enhanced customer satisfaction levels, manufacturers must provide the highest-quality products to their customers
Reading bar codes	In manufacturing, the role of bar code scanners is truly valuable. Bar code scanners have replaced the traditional paper-and-pencil method of processing
Improving worker safety	Providing improved health to workers and keeping the environment safe is of prime concern for manufacturers

Industrial Revolutions

- **Industry 1.0** → 1784 → Mechanisation and use stream of water power
- **Industry 2.0** → 1870 → Mass production using electrical power
- **Industry 3.0** → 969 → Automated production. It is based on systems
- **Industry 4.0** → Today → Smart factory, Autonomous systems

Enabling Technologies

- Big data
- IoT, IIoT and Cloud
- ML and DL
- Augmented reality
- Cobot
- Cyber-physical systems

Practical Challenges

- Lack of access to proof points
- Require large-scale investment
- Shortage of collaboration
- Security and privacy concerns
- Not an apparent strategic plan
- Be deficient in standards
- Immature data analysis

Main Activities

- Inbound logistics
- Operations
- Outbound operations
- Marketing
- Sales
- Service

Machine Vision Applications

- Absence/presence detection
- Automated vision testing and measurement
- Bar code reading
- Color verification
- Defect detection
- Vision-guided robots
- Optical character recognition and verification
- Part verification
- Pattern matching
- Sorting
- Traceability

Goals That Machine Vision Systems Can Help Achieve Include the Following:

Higher quality → Inspection, measurement, gauging and assembly verification

Increased productivity → Repetitive tasks formerly done manually are now done by Machine Vision System

Production flexibility → Measurement and gauging/Robot guidance/Prior operation verification

Less machine downtime and reduced setup time → Changeovers programmed in advance

More complete information and tighter process control → Manual tasks can now provide computer data feedback

Lower capital equipment costs → Adding vision to a machine improves its performance, avoids obsolescence

Lower production costs → One vision system vs. many people/Detection of flaws early in the process

Scrap rate reduction → Inspection, measurement and gauging

Inventory control → Optical character recognition and identification

Reduced floor space → Vision system vs. operator

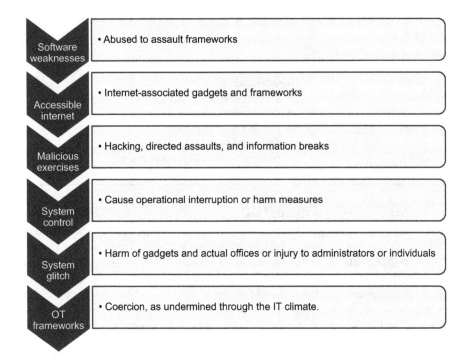

FIGURE 4.2 Security hazards in IIoT.

Ensuring IIoT systems in this manner requires related peril protection and beginning-to-end protection, from the entryway to the endpoint that can give the following:

- Regular checking and disclosure if there ought to emerge an event of malware sickness
- Better threat detectable quality and early disclosure of irregularities
- Proactive counteraction of threats and attacks among IT and OT
- Secure information move
- A cutting-edge IPS to prevent assaults from misusing weaknesses
- Server and application insurance across the server farm and the cloud

Smart Manufacturing: Smart manufacturing permits plant administrators to gather and investigate information to settle on better educated choices and upgrade creation. Table 4.3 shows the approaches and characteristics that are related to the smart manufacturing.

- The information from sensors and machines is imparted to the cloud by IoT network arrangements sent at the production line level.
- This information is investigated and joined with logical data and afterward imparted to approve partners.

TABLE 4.3
Smart Manufacturing

Approach	Characteristics
Market-Driven	• Used in private sector management
	• Depends on the global policy
Coordinated	• Used in government sectors
	• Some restrictions
Manageable	• Used both private and government sectors
	• Depends on the investment and new technology

IoT innovation, utilising both wired and remote availability, empowers this progression of information and gives the capacity to screen and oversee measures distantly and change creation designs rapidly, progressively when required.

Smart Metering: Smart metering profit utilities by improving consumer constancy with faster cooperation, while bountiful customers better control their energy use to set aside cash and lessen fossil fuel byproducts (H. Ahuett-Garza et al. 2018). With power permeability right to the meter, utilities can advance energy conveyance and even build a movement to move request many. Smart metering encourages utilities to (P. Lade et al. 2017):

• Reduce working costs by overseeing manual tasks distantly
• Improve determining and smooth out force utilisation
• Improve client care through profiling and division
• Reduce energy burglary
• Simplify miniature age checking and track sustainable force

Also, smart meters can empower a variety of new smart administrations to private, business and industrial clients to all the more likely deal with their energy utilisation designs, decrease in general force utilisation and even advantage from new valuing models (S. Jeschke et al. 2017).

Smart Agriculture: With the management of IIoT in agriculture, unmistakably further developed sensors are being used (L. Banjanovic-Mehmedovi´c et al. 2020). The sensors are presently connected with the cloud using cell/satellite organisation. Which lets us know the continuing in sequence from the sensors, settling on dynamic viability? The utilisation of IoT in the agriculture business has assisted the ranchers with observing the water tank points endlessly, which makes the water system measure more proficient. The progression of IoT innovation in agriculture tasks has gotten the utilisation of sensors each progression of the cultivating cycle like what amount of time and assets a seed requires to turn into a developed vegetable. Figure 4.3 shows the supporting methods available by IIoT in smart agriculture.

IoT in agriculture has come up as a second surge of green agitation. The focal points that the farmers are getting by changing IoT are twofold (S. Sudha et al. 2016). It has helped farmers with a decrease in their costs and addition yields all the while by improving farmer's dynamic with exact information.

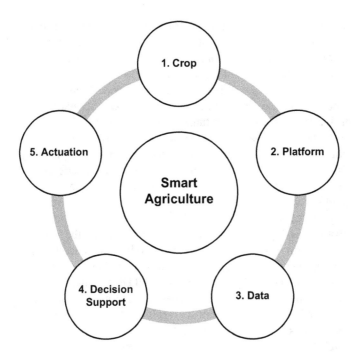

FIGURE 4.3   Smart agriculture–supporting methods.

4.4   CONSEQUENCES OF THE EFFECTIVE DESIGN WHILE USING DEEP LEARNING–IIoT

We have examined the significance of deep learning (DL) in different ventures; nonetheless, its effective usage in the modern climate and obtaining trustful and helpful outcomes are difficult. It requires domain understanding with a critical thinking mentality and individuals with measurable investigation foundations who are acceptable at discovering important experiences into the information (H. Yan et al. 2018). In accompanying, some of the consequences of the effective design while using DL-IIoT is listed in Figure 4.4.

Complication: Complication is one of the significant issues of DL models, which requires an additional push to address. One of the issues concerning DL execution is the time-utilisation of the training stage and calculation necessity because of the intricacy of the model and huge mechanical dataset (F. Liang et al. 2020). The subsequent issue is the absence of an enormous number of training tests in modern situations, which diminishes the exactness and productivity of models by overfitting (C. Qiu et al. 2018).

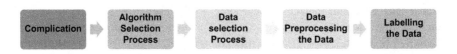

FIGURE 4.4   Consequences of the effective design while using deep learning IIoT.

Algorithm Selection Process: There are a few generally available famous DL calculations to be utilised in IIoT applications. These calculations can work in any nonexclusive situation; however, choosing a calculation for explicit modern applications should be founded on specific rules available. For every calculation, it is fundamental to know which DL calculation will give its best in which situation (J. Wang et al. 2018). The inappropriate choice of a DL calculation can cause numerous issues like the creation of trash yields, which prompts an exercise in futility, exertion and cash (X. Ji et al. 2018).

Data Selection Process: A DL algorithm's prosperity is straightforwardly connected with the preparation information, which is all around depicted by a renowned saying, "Information in, Data out." The right kind and measure of information are basic for any DL algorithm. It is fundamental to dodge such information that causes specific predisposition and make or select the information illustrative of the case in a modern cycle (X. Ma et al. 2018).

Data Preprocessing the Data: In the wake of choosing the information or data, the following significant advance would be to convert the chaotic information containing missing qualities, exceptions and useless sections to the type of information that could be perceived by measurements and DL calculations. This progression incorporates parsing, cleaning and preprocessing information like changing over the other type of information into numbers, scaling of highlights and eliminating or supplanting missing sections (P. Li et al. 2017).

Labeling the Data: Information marking or data labeling for administered DL calculations is testing and can't be re-appropriated for cutting-edge and escalated assignments. Marking of clinical imaging on which a characterisation model could be prepared for determination measure needs area specialists. Notwithstanding, the issue is that particular clinical specialists see this movement as a tedious cycle (T. J. Saleem et al. 2019). Other than these basic difficulties, numerous different difficulties exist in the field of DL in a keen modern climate, including overseeing model adaptations, information variant, duplicating the models.

4.5 VARIOUS DEEP LEARNING TECHNIQUES USED IN INDUSTRIES

DL strategies have made progress in tackling automatic acknowledgment of examples in information, which outperformed people. In the most recent decade, DL tackled the restrictions of customary AI calculations. During its development period, it grabbed the attention of organisations and everybody wants to utilise it. However, in any event, for profoundly prepared experts, it is difficult to investigate and characterise where to begin and actualise a DL model to tackle their issues.

Convolution Neural Networks (CNNs): A network layer that utilises the convolution activity is alluded to as a convolutional layer. Significantly, CNN is applied to two-dimensional sensor information. CNNs can likewise be applied to one-dimensional sensor information (D. Weimer et al. 2015). There are two sections in CNN engineering. The initial segment comprises convolutional and pooling layers that are shown in Figure 4.5 which assist with extricating highlights from the information, and the subsequent part learns a portrayal of the preparation information to foresee the objective variable (A. Essien et al. 2020).

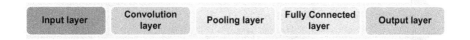

FIGURE 4.5 CNN layer format.

Recurrent Neural Networks (RNNs): The fundamental idea driving the RNN is that it utilises successive data where the yield of the past advance is taken care of alongside the contribution of the current advance (A. Sherstinsky 2020). So, it is a progression of data sources reliant on one another to infer the last yield, not at all like the conventional neural networks where sources of information and yields are autonomous of one another, which is shown in Figure 4.6. For instance, the previous sentence has importance since words are orchestrated in a particular way; on the off chance that we move words out of their set request, the significance will change (B. Roy et al. 2018). The succession matters. RNN has discovered incredible accomplishment in conveying with regards to natural language processing (Q. Wu et al. 2018). Long short-term memory (LSTM) is the most regularly utilised RNNs, and it is viewed as better at catching long-haul conditions. LSTMs are practically equivalent to RNN. In any case, we won't dig into LSTM until further notice. Allow us to examine what RNN can achieve (X.-B. Jin et al. 2020).

Autoencoders: An autoencoder is a special neural network to learn a portrayal that intently imitates the contributions as the yield (L. Ren et al. 2018). The autoencoder has two sections, encoder and decoder, which is shown in Figure 4.7. The information and yield layers have a similar number of hubs and all the layers are completely associated; however, the network acquaints a bottleneck with powering the network to learn just the fundamental highlights (L. Wen et al. 2017).

The bottleneck is presented by picking a number of hubs in the interfacing layer, between the encoder part and the decoder part, to be not exactly the number in the information layer. The preparation cycle is like other neural networks, where the loads and inclination of the network are found out while limiting a misfortune work (Y. Meidan et al. 2018). The encoder learns a portrayal of the information and the decoder reproduces the contribution from the educated portrayal. This cycle of learning and recreation is utilised for different purposes. Denoising and dimensionality decrease are applications that are pertinent. Denoising can help in improving the nature of information and consequently the nature of the conjectures. Dimensionality decrease can help improve computational productivity when the information is high dimensional.

FIGURE 4.6 RNN layer format.

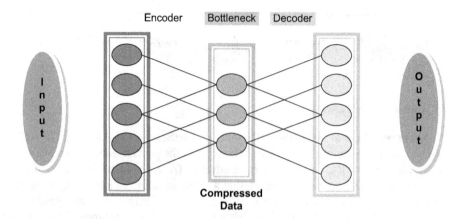

FIGURE 4.7   Autoencoder format.

4.6   CURRENT STATUS OF DEEP LEARNING IN IIoTS

In the primary industrial insurgency, steam and automation tested mechanical architects. These moves are based on tackling steam to make machines that delivered products on the scale. In the subsequent upset, the appearance of power and the 24-hour manufacturing plant tested prevalently operational and electrical designing (M. Liu et al. 2019). For operational administration, the challenge was the production of the advanced processing plant culture that worked on move work and required huge authoritative changes not already conceivable when work was generally "daybreak to sunset."

With the IIoT in the current insurgency, all regions of activities are tested. By making a production line of digital actual frameworks, mechanical, electrical and operational specialists are compelled to break out of direct assumptions for improving productivity just inside their circle (M. A. Al-Garadi et al. 2020). Furthermore, with the appearance of the associated plant, information has detonated to a level that would have recently been unmanageable. The new truth is that advancements, for example, AI, DL and AI, necessitate that operational productivity is seen as a solitary substance influencing the whole activity and where all utilitarian zones add to upgrades as a component of a coordinated and associated framework that augments effectiveness (L. Zeng et al. 2019).

Operational Efficiency and the Limits: The most recent years have likewise seen the ascent of cycle and quality improvement procedures pointed toward expanding operational effectiveness. From multiple points of view, these systems were an endeavor to unite information and measurements accessible through the more complex instruments that the third industrial transformation could convey. Every one of these frameworks and their philosophies were exceptionally effective and took into consideration colossal additions for organisations that received them. What's more, their exhaustive hierarchical structure affected the whole association. Accordingly, they speak to the primary endeavors to oversee the creation and drive measure improvement and productivity comprehensively across a plant or a whole organisation (Y. Chen 2017).

Yet, while these projects were effective in Industry 4.0, there are still impediments to how far they can go as a framework. To begin with, regardless of dependence upon measurements and the utilisation of PCs and progressed strategies for examination, they are as yet human-driven frameworks. Projects should be driven, advocated, sharpened and authorised inconclusively (M. Ranzato et al. 2008). Furthermore, obstructions, for example, culture, exhaustion, burnout and interdepartmental rivalry fortify the truth that there is just so much human mediation that can do to impact the framework physically. Besides, there is simply the issue of the information. Once more, while progressed scientific strategies are utilised in the two frameworks, they are frequently still silted and the choices that come from them are as yet made by people. Improving operational effectiveness further will require the dependence on the IIoT inside "shrewd"-associated manufacturing plants (A. Hassanzadeh et al. 2016). Here, measure improvement is driven not by human activities or straight floods inside explicit fields, rather by the total joining of PC and machine that envelops the whole activity. This is conceivable through the presentation of AI, and deep investigation programming that can help understand a few advantages:

- These advancements can perform at a miniature level that people can't.
- They can work quicker than people.
- They can take into consideration quicker, evenhanded and more precise dynamics.
- Systems can measure and investigate information to "see" examples and patterns not promptly discernable by people.
- They can decentralise dynamic by permitting numerous self-governing or semi-self-governing choices from with the stages themselves.

4.6.1 OBSTRUCTIONS TO INDUSTRIAL IIoT ADOPTION

Before assembling organisations that can exploit what IIoT has to bring to the table, there are a couple of outstanding obstructions that should be survived.

Data Interoperability: One such obstruction is that of information inoperability. Creation, planning, quality, designing and business checking to the program may not be connected, or, on the off chance that they are connected, may at present not be viable for investigation (D. S. Evans 2009).

IT Skill Sets: Another hindrance to utilising IIoT potential for improved operational effectiveness is the absence of abilities and admittance to aptitudes that will be required for sending. As another and advancing innovation, numerous organisations have not started to redesign these aptitudes and may not completely comprehend that they need to or what is needed to do as such (Z. Liang et al. 2014).

Security and Privacy: Perhaps probably the greatest boundary to faster appropriation is security. The same number of gadgets is controlled by Open-Source Programming (OSP), and new associations leave open the opportunities for a break. Furthermore, as a rule, the OSP programming isn't screened for weaknesses. Aggravating the issue is the truth that there is no limit to end security arrangement accessible, leaving makers and specialist co-ops at chances over who ought to give security and how that should be refined (J. Pan et al. 2017).

4.7   DEEP LEARNING REVOLUTION IN IIoTs

DL is a division of ML, which together plunge into AI and IoT. Exclusive of DL appliances, however, robotisation and intelligence would not be the place where they are today (G. Dimitrakopoulos et al. 2010). For example, AI in assembling has progressed significantly with technology like prescient protection. Nonetheless, this industry isn't just one profiting from a DL application.

Digital Supporting Devices: Perhaps the most well-known DL applications are digital supporting devices. They are becoming more standard in the present society. Your cell phones and gadgets undoubtedly have one from a major tech organisation like Cortana, Alexa, or Google. These AI manifestations utilise normal natural programming language preparing to understand and do your solicitations. Something as basic as requesting that Alexa play music utilises DL. Another model happens when Siri adjusts to your examples and inclinations.

Energy: The energy business is continually changeable. Regardless of whether it is the discussion about sustainable power or the changing costs of oil and gas, it's a broad field. Laborers in this industry, however, have been adjusting to innovation. With the developing populace comes more demand for energy and power. Industry laborers can utilise tech with DL capacities to change their creation standards dependent on the information they receive. Maintenance and checking, as well, require difficult work. DL applications like prescient maintenance and infrared tech make it all simpler.

Agriculture: Agriculture is everywhere, even in places like urban communities where it is less evident. In any case, since it's an industry that works with nature doesn't mean it's not utilising the most recent progressions (N. Suma et al. 2017). With an Earth-wide temperature boost comes a decrease in certain harvest yields. This variance expects laborers to think ahead concerning how they will maintain organic market necessities. DL expectation is one of the arrangements. Laborers use information from satellites and sensors to understand the coming climate designs better. At that point, they get a kick-off on the creation and guarantee that temperatures don't ruin their yields.

Industrialised: Perhaps the most essential business on the planet is producing. Without it, organisations would not have the option to sell their items, and the store network would presently don't exist. There's a plenitude of it, as well. Prescient maintenance is available in different enterprises, yet fabricating is the place where it flourishes. Hardware and machines are characteristic pieces of the business. Understanding when to fix them before they break will save time, cash and energy. Different instances of DL-AI in assembling incorporate deal determining and progressed examination. With these devices, producers can perceive how high or low market interest levels will reach in the coming months. At that point, they can change their arrangements in like manner.

Trading: After assembling, retail is the last stop of the inventory network. Anybody in this industry endeavors to make their items and administrations stand apart from clients. Fortunately, profound learning applications help that cycle along. With every exchange and snapshot of commitment regardless of whether coming up or on the Web information is the result. For example, profound learning calculations gain proficiency with an online client's inclinations and purchasing propensities. The framework would then be able to recommend customised content that will coordinate their inclinations. It's an unobtrusive cycle that goes far.

Food: Much like retail and horticulture, adjusting to innovation is the path forward. For instance, 2020 was the time of food conveyance applications. Individuals are remaining at home like never before, and the food business has changed as a result of it. Food organisations can utilise profound figuring out how to comprehend current shopper conduct. At that point, it can anticipate future patterns. From a rural viewpoint, different pieces of the food business can utilise information calculations to examine significant patterns. From that point, the inventory network can cooperate to fulfill the evolving needs.

Cybersecurity: As IoT alters the world through innovative gadgets and Web associations, it likewise opens up more technology to weaknesses. Cybercriminals utilise unstable organisations to assault and break different gadgets regardless of whether it's an individual telephone or an organisation's information base (S. Jeschke et al. 2017). Here, a profound learning application implies that organisations and people remain one stride in front of the assaults. Computer-based intelligence-based programming that ensures IoT gadgets is the new standard. Information is too significant to even think about losing, and profound learning can help.

Automobile Industry: The automobile business bays a wide scope of vehicles. It's additionally one of the clearest places to search for AI advancement, for example, self-driving vehicles. Tesla is a notable model in the automobile industry. These vehicles make them drive alternatives and progressed location controls. For example, in case you're struggling to zero in out and about, the vehicle's neural organisations help with direction and fixation. Its transformation to your inclinations and propensities is the place where profound learning shows its actual force.

Health Management: The last urgent profound learning application is in the medical services industry. Here, this type of knowledge can be lifesaving. Diagnosing and treating is basically where profound learning makes a difference. As new calculations create, they utilise profound figuring out how to investigate patterns, examples and practices. The result is prescient diagnostics, where the framework may foresee which patients may get sick. On the treatment side, medical care laborers can utilise similar information and calculations to customise care. Obliging the necessities of the patient will take the business forward.

4.8 CASE STUDY: INTEGRATING DEEP LEARNING AND INDUSTRIAL IoT IN THE FUTURE ERA

DL-based IIoT gadgets can upset conventional on-location work capacities through far-off working and virtual gatherings. DL has a crucial job in quickening Industry 4.0 appropriation, through the combination of quicker correspondence innovation with IIoT gadgets that are now utilised for an assortment of industrial tasks, including support, computerisation and cobots, distant gatherings and inventory network the executives. Be that as it may, DL is viewed as a need to improve the inactivity and volume of information.

DL as an empowering agent of the ongoing investigation of industrial tasks increased or augmented reality gatherings associated with independent manufacturing plants and remote, agreeable advanced mechanics. Fusing DL in IIoT gadgets will empower low inactivity, increment information throughput and diminish activity time, consequently prompting improved, generally speaking, cycle efficiency.

"Aside from improving the mechanisation of industrial activities and control, Deep Learning–empowered IIoT gadgets can likewise limit the unpredictability of inventory network organisations and distribution center administration, assisting organisations with effectively working in powerful business conditions." Not all industrial activities require DL, indeed, numerous presently depend on Bluetooth and others. For activities like windmill ranches and offshore oil drilling, the expense may restrict advancements and DL for one more decade.

Monetary administrations and protection are key areas for DL-incorporated IIoT development. Mechanised DL frameworks can perform every day, tedious assignments consistently while diminishing human mistakes simultaneously. DL may likewise empower self-driving vehicles by moving a portion of the calculation outside the vehicle. The vehicle would likewise have the option to collaborate with an associated traffic framework, giving constant information on mishaps and street conditions. The current COVID-19 pandemic has resulted in many advances, to conspicuousness. Presently, accept that DL will be key for business supportability: "The low inertness will help in dealing with the high traffic to Web-based business by improving organisation availability at a quicker speed, quickening on the Web buys and request positions."

4.9 CONCLUSION

This chapter presented the perspectives on DL techniques for IIoT with different angles that include data analysis and visualisation for the ML process, IIoT in the form of smart manufacturing, smart metering and smart agriculture. This study also illustrates the consequences of the effective design while using DL IIoT through the algorithm selection process, data selection process, data preprocessing the data and labeling the data. Furthermore, the various DL techniques and their uses in different industries focus on CNN, RNN and autoencoders. The current status of DL in IIoTs also has been discussed with the subtopics on operational efficiency and the limits that include data interoperability, skill sets, security and privacy. Finally, the DL revolution in IIoT has been elaborated in terms of digital supporting devices, energy, agriculture, industrialisation, trading, food, cybersecurity, automobile industry and health management. The case study of integrating DL and IIoT in the future era also has been interpreted. From the investigations, it is presumed that DL utilises various methodologies for tackling trainable numerical models to dissect information, and for all the features ML gets, it is additionally just one of a wide range of strategies available for cross-examining information— and not the most ideal alternative.

REFERENCES

H. Ahuett-Garza and T. Kurfess. 2018. A Brief Discussion on the Trends of Habilitating Technologies for Industry 4.0 and Smart Manufacturing. *Manuf. Lett.* 15. 60–63.

M. A. Al-Garadi, A. Mohamed, A. Al-Ali, X. Du, I. Ali, and M. Guizani. 2020. A Survey of Machine and Deep Learning Methods for the Internet of Things (IoT) Security. *IEEE Commun. Surv. Tutorial.* 22(3). 1646–1685

K. Alrawashdeh and C. Purdy. 2016. Toward an Online Anomaly Intrusion Detection System based on Deep Learning. *IEEE Int. Conf. on Machine Learning and Appl.* 195–200.

L. Banjanovic-Mehmedovi´c and F. Mehmedovi. 2020. Intelligent Manufacturing Systems Driven by Artificial Intelligence in Industry 4.0. In *Handbook of Research on Integrating Industry 4.0 in Business and Manufacturing.* IGI Global. 31–52.

F. P. Carvalho. 2017. Mining Industry and Sustainable Development: Time for Change. *Food Energy Secur.* 6(2). 61–77.

B. Chen and J. Wan. 2019. Emerging Trends of ML-based Intelligent Services for the Industrial Internet of Things (IIoT). *IEEE Computer Comm. and IoT Appl. (ComComAp).* 135–139.

Y. Chen. 2017. Integrated and Intelligent Manufacturing: Perspectives and Enablers. *Engineering.* 3(5). 588–595.

L. Da Xu, W. He, and S. Li. 2014. Internet of Things in Industries: A Survey. *IEEE Trans. Ind. Inf.* 10(4). 2233–2243.

G. Dimitrakopoulos and P. Demestichas. 2010. Intelligent Transportation Systems. *IEEE Veh. Tech. Mag.* 5(1). 77–84.

A. Essien and C. Giannetti. 2020. A Deep Learning Model for Smart Manufacturing using Convolutional LSTM Neural Network Autoencoders. *IEEE Trans. Ind. Inf.* 16(9). 6069–6078.

D. S. Evans. 2009. The Online Advertising Industry: Economics, Evolution, and Privacy. *J. Econ. Perspect.* 23(3). 37–60.

J. Gubbi, R. Buyya, S. Marusic, and M. Palaniswami. 2013. Internet of Things. 2013. A Vision, Architectural Elements, and Future Directions. *Future Gener. Comp. Syst.* 29(7). 1645–1660.

Y. Han, C.-J. Zhang, L. Wang, and Y.-C. Zhang. 2019. Industrial IoT for Intelligent Steelmaking with Converter Mouth Flame Spectrum Information Processed by Deep Learning. *IEEE Trans. Ind. Inf.* 16(4). 2640–2650.

A. Hassanzadeh, S. Modi, and S. Mulchandani. 2015. Towards Effective Security Control Assignment in the Industrial Internet of Things. *IEEE World Forum on the Internet of Things (WF-IoT).* 795–800.

S. Jeschke, C. Brecher, T. Meisen, D. Ozdemir, and T. Eschert. 2017. Industrial Internet of Things and Cyber Manufacturing Systems. In *Industrial Internet of Things.* Springer, Cham. 3–19.

X. Ji, Y. Jiang. 2018. Deep Learning for Smart Agriculture: Concepts, Tools, Applications, and opportunities. *Int. J. Agric. Biol. Eng.* 11(4). 32–44.

X.-B. Jin, N.-X. Yang, X.-Y. Wang, Y.-T. Bai, T.-L. Su, and J.-L. Kong. 2020. Hybrid Deep Learning Predictor for Smart Agriculture Sensing based on Empirical Mode Decomposition and Gated Recurrent Unit Group Model. *Sensors.* 20(5). 1334–1339.

R. A. Khalil, E. Jones, M. I. Babar, T. Jan, M. H. Zafar, and T. Al-Hussain. 2019. Speech Emotion Recognition Using Deep Learning Techniques: A Review. *IEEE Access.* 7(117). 327–345.

P. Lade, R. Ghosh, and S. Srinivasan. 2017. *Manufacturing Analytics and Industrial Internet of Things. IEEE Intell. Syst.* 32(3).74–79.

Y. LeCun, Y. Bengio, and G. Hinton. 2015. Deep Learning. *Nature*. 521(7553). 436–444.

K. Lepenioti, M. Pertselakis, A. Bousdekis, A. Louca, F. Lampathaki, D. Apostolou, G. Mentzas, and S. Anastasiou. 2020. Machine Learning for Predictive and Prescriptive Analytics of Operational Data in Smart Manufacturing. *Int. Conf. on Adv. Information. Syst. Engg.* Springer. 5–16.

B.-H. Li, B.-C. Hou, W.-T. Yu, X.-B. Lu, and C.-W. Yang. 2017. Applications of Artificial Intelligence in Intelligent Manufacturing: A Review. *Front. Inf. Technol. Electron. Eng.* 18(1). 86–96.

P. Li, Z. Chen, L. T. Yang, Q. Zhang, and M. J. Deen. 2017. Deep Convolutional Computation Model for Feature Learning on Big Data in the Internet of Things. *IEEE Trans. Ind. Inf.* 14(2). 790–798.

F. Liang, W. Yu, X. Liu, D. Griffith, and N. Golmie. 2020. Toward Edge Based Deep Learning in Industrial Internet of Things. *IEEE Internet Things J.* 7(5). 4329–4341.

Z. Liang, G. Zhang, J. X. Huang, and Q. V. Hu. 2014. Deep Learning for Healthcare Decision Making with *EMRs. IEEE Int. Conf. on Bioinformatics and Biomedicine (BIBM).* 556–559.

M. Liu, F. R. Yu, Y. Teng, V. C. Leung, and M. Song. 2019. Performance Optimization for Blockchain-enabled Industrial Internet of Things (IIoT) systems: A Deep Reinforcement Learning Approach. *IEEE Trans. Ind. Inf.* 15(6). 3559–3570.

X. Ma, T. Yao, M. Hu, Y. Dong, W. Liu, F. Wang, and J. Liu. 2018. A Survey on Deep Learning Empowered IoT Applications. *IEEE Access.* 7(181). 721–732.

M. S. Mahdavinejad, M. Rezvan, M. Barekatain, P. Adibi, P. Barnaghi, and A. P. Sheth. 2018. Machine Learning for the Internet of Things Data Analysis: A Survey. *Digital Commun. Networking.* 4(3). 161–175.

N. Mehdiyev, J. Lahann, A. Emrich, D. Enke, P. Fettke, and P. Loos. 2017. Time Series Classification using Deep Learning for Process Planning: A Case from the Process Industry. *Procedia Comput. Sci.* 114. 242–249.

Y. Meidan, M. Bohadana, Y. Mathov, Y. Mirsky, A. Shabtai, D. Breitenbacher, and Y. Elovici. 2018. N-bait? Network-Based Detection of IoT Botnet Attacks Using Deep Autoencoders. *IEEE Pervasive Comput.* 17(3). 12–22.

A.-H. Muna, N. Moustafa, and E. Sitnikova. 2018. Identification of Malicious Activities in Industrial Internet of Things Based on Deep Learning Models. *J. Inf. Secur. Appl.* 41. 1–11.

M. M. Najafabadi, F. Villanustre, T. M. Khoshgoftaar, N. Seiya, R. Wald, and E. Muharemagic. 2015. Deep Learning Applications and Challenges in Big Data Analytics. *J. Big Data.* 2(1). 1–7.

J. Pan and J. McElhannon. 2017. Future Edge Cloud and Edge Computing for the Internet of Things Applications. *IEEE Internet Things J.* 5(1). 439–449.

S. Prathibha, A. Hong, and M. Jyothi. 2017. IoT-Based Monitoring System in Smart Agriculture. *IEEE Int. Conf. on Recent Adv. in Electr. & Comm. Techn. (ICRAECT).* 81–84.

M. K. Putchala. 2017. Deep Learning Approach for Intrusion Detection System IDS) in the Internet of Things (IoT) Network Using Gated Recurrent Neural Networks (GRN).

C. Qiu, F. R. Yu, H. Yao, C. Jiang, F. Xu, and C. Zhao. 2018. Blockchain-Based Software-Defined Industrial Internet of Things: A Dueling Deep Q-Learning approach. *IEEE Internet Things.* 6(3). 4627–4639.

M. Ranzato, Y.-L. Boureau, and Y. L. Cun. 2008. Sparse Feature Learning for Deep Belief Networks. *Adv. Neural Information Process Systems.* 1185–1192.

L. Ren, Y. Sun, J. Cui, and L. Zhang. 2018. Bearing Remaining Useful Life Prediction Based on Deep Auto encoder and Deep Neural Networks. *J. Manuf. Syst..* 48. 71–77.

B. Roy and H. Cheung. 2018. A Deep Learning Approach for Intrusion Detection in the Internet of Things Using Bi-Directional Long Short-Term Memory Recurrent Neural Network. *IEEE Int. Telecomm. Network and Appl. Conf. (ITNAC)*. 1–6.

T. J. Saleem and M. A. Chishti. 2019. Deep Learning for the Internet of Things Data Analytics. *Proc. Comp. Sci.* 163. 381–390.

A. Sherstinsky. 2020. Fundamentals of Recurrent Neural Network (RNN) and Long Short-Term Memory (LSTM) Network. *Physica D*. 404. 132306–132309.

E. Sisinni, A. Saifullah, S. Han, U. Jennehag, and M. Gidlund. 2018. Industrial Internet of Things: Challenges, Opportunities, and Directions. *IEEE Trans. Ind. Inf.* 14(11). 4724–4734.

S. Sudha, M. Vidhyalakshmi, K. Pavithra, K. Sangeetha, and V. Swathi. 2016. An Automatic Classification Method for Environment: Friendly Waste Segregation using Deep Learning. *IEEE Technological Innovations in ICT for Agriculture and Rural Development (TIAR)*. 65–70.

N. Suma, S. R. Samson, S. Saranya, G. Shanmugapriya, and R. Subhashri. 2017. IoT-Based Smart Agriculture Monitoring System. *Int. J. Recent Innov. Trends Comput. Commun.* 5(2). 177–181.

A. Thantharate, R. Paropkari, V. Walunj, and C. Beard. 2019. DeepSlice: A Deep Learning Approach towards an Efficient and Reliable Network Slicing in 5G Networks. *IEEE Annual Ubiquitous Computer Electr. & Mobile Comm. Conf. (UEMCON)*. 0762–0767.

F. Ullah, H. Naeem, S. Jabbar, S. Khalid, M. A. Latif, F. Al-Tudjman, and L. Mostarda. 2019. Cyber Security Threats Detection in the Internet of Things Using Deep Learning Approach. *IEEE Access*. 7(124). 379–389.

J. Wang, Y. Ma, L. Zhang, R. X. Gao, and D. Wu. 2018. Deep Learning for Smart Manufacturing: Methods and Applications. *J. Manuf. Syst.* 48. 44–156.

D. Weimer, A. Y. Benggolo, and M. Freitag. 2015. Context-Aware Deep Convolutional Neural Networks for Industrial Inspection. *Proc. of the Australian Conf. on Artificial Intelligence*, Canberra, Australia. 30.

L. Wen, L. Gao, and X. Li. 2017. A New Deep Transfer Learning based on Sparse Auto-Encoder for Fault Diagnosis. *IEEE Trans. Syst. Man Cybern. Syst.* 49(1). 136–144.

T. Wong and Z. Luo. 2018. Recurrent Auto-Encoder Model for Large-Scale Industrial Sensor Signal Analysis. *Int. Conf. on Engg. App. Of Neural Network*. Springer. 203–216.

Q. Wu, K. Ding, and B. Huang. 2018. Approach for Fault Prognosis Using Recurrent Neural Network. *J. Intell. Manuf.* 31(3). 1–13.

Y. Xu, S. Li, D. Zhang, Y. Jin, F. Zhang, N. Li, and H. Li. 2018. Identification Framework for Cracks on a Steel Structure Surface by a Restricted Boltzmann Machines Algorithm based on Consumer-grade Camera Images. *Struct. Control Health Monit.* 25(2). 1–12.

H. Yan, J. Wan, C. Zhang, S. Tang, Q. Hua, and Z. Wang. 2018. Industrial Big Data Analytics for Prediction of Remaining Useful Life Based on Deep Learning. *IEEE Access*. 6(17). 190–197.

F. Zantalis, G. Koulouras, S. Karabetsos, and D. Kandris. 2019. A Review of Machine Learning and IoT in Smart Transportation. *Future Internet*. 11(4). 94–97.

L. Zeng, E. Li, Z. Zhou, and X. Chen. 2019. Boomerang: On-Demand Cooperative Deep Neural Network Inference for Edge Intelligence on the Industrial Internet of Things. *IEEE Networks*. 33(5). 96–103.

Q. Zhang, L. T. Yang, Z. Chen, P. Li, and F. Bu. 2018. An Adaptive Dropout Deep Computation Model for Industrial IoT Big Data Learning with Crowdsourcing to Cloud Computing. *IEEE Trans. Ind. Inf.* 15(4). 2330–2337.

P. Zheng, Z. Sang, R. Y. Zhong, Y. Liu, C. Liu, K. Mubarok, S. Yu, X. Xu. 2018. Smart Manufacturing Systems for Industry 4.0: Conceptual Framework, Scenarios, and Future Perspectives. *Front. Mech. Eng.* 13(2). 137–150.

5 Proposal for Missing Person Locator and Identifier Using Artificial Intelligence and Supercomputing Techniques

Edgar Gonzalo Cossío Franco, Yotziri Paloma Pérez Rios, Ramiro Aguilar Ordaz, Iván Alberto Cruz García, Kevin Gálvez Parra and Héctor Manuel Gutiérrez Zazueta

CONTENTS

DOI: 10.1201/9781003122401-5

5.1   INTRODUCTION

Facial recognition has become one of the most discussed technological solutions in recent years.

Several companies have begun to develop computer algorithms that are capable to perceive live images and video of people's faces for identification, captured through camera networks.

Although other types of biometric systems, such as iris scanning, are at best much more accurate than facial recognition (with relatively low error rate; just under 1%), facial recognition is one of the most widely accepted biometric systems. Moreover, it does not require the user to tap, insert or perform any action such as clicking on anything (Monroe, 2009).

On one side, there are defenders who highlight the technology capacity to provide greater comfort and security to users. Meanwhile, critics tend to point out this tool as nonfunctional or, on the contrary, as an invasion of privacy and, therefore, a threat to society. Some implementation concerns of such technology are valid; however, technological advancements have shown the highly scalable potential to take advantage on the security aspects, mainly in cases to search missing people.

The current process to solve the cases of missing people is that citizens and authorities can look out by themselves. In some cases, missing people are found because someone has noticed them in their own surroundings; however, only a small portion of cases are resolved in this way, even though there is a wide coverage of cases of missing people. There are many reasons why citizens may not be able to help in locating missing people, including the likelihood of finding, noticing and recognising that missing person.

For a better understanding of the finding and recognising people process, intelligent systems can be built to increase the likelihood of expediting this search and recognition process and, therefore, facilitate the recovery of people on this situation.

The usage of facial recognition techniques (Peace, 2018) makes it easier to find missing people, and this directly addresses all the disadvantages of using other conventional means.

Some platforms such as social media can be used to find missing people; sometimes this method is not the best, and issues come out, for example, sometimes people don't want to be found, and this advertising makes them to walk away because they don't want to be found, and this is one of the highest risk.

As a result of this problem, a facial recognition system is proposed for the identification of missing people, which is maintained by tracking through live video transmission on a network of security cameras, using a data set of people who are in this delicate situation, which is provided by their families. The system has a notification module that sends an alert to the appropriate authorities to intercept the perpetrator identified by the system in real time.

5.1.1 ARTIFICIAL INTELLIGENCE

Artificial intelligence (AI) is defined as the discipline of computer science that makes the computers exhibit intelligent or human-like behavior. AI also refers to computer systems that exhibit complex behavior similar to live systems, such as swarms, ant colonies, microbiology and neural systems.

There is a distinction between "strong" AI—having computational functions with strong similarities to human thinking, showing some self-awareness—on the other hand, we define "weak" AI—the computational applications of which deal with limited application areas and contain a range of practical knowledge and appear to have some intelligent characteristics—in this case, we can mention expert systems and heuristic search algorithms (Chowdhury et al., 2012).

5.1.2 SMART CITIES

To provide services that offer more efficient alternatives in security aspects, Mexico should consider increasing the usage of AI in the creation and expansion of smart cities with the intention to generate safer environments. The term smart city (SC) has been used in different perspectives, and the following are some descriptions of an SC:

1. SCs are often defined as virtually reconstructed cities. The term has been used interchangeably as equivalent to *digital city, information city, wired city, tele-city, knowledge-based city, electronic communities, electronic community spaces, flexicity, teletopia, cybervillage*, encompassing a wide range of electronic and digital applications related to digital spaces in communities and cities (Droege, 1997).
2. The World Foundation for Smart Communities gives another meaning, linking digital cities with smart growth, a development based on information and communication technologies.
3. A smart community is a community that has made a conscious effort to use the technology information to transform life and work in its region in meaningful and fundamental ways, rather than incrementally (California Institute for Smart Communities [CISC], 2001).
4. SCs are defined as environments with integrated information and communication technologies that create interactive spaces that bring computing into the physical world. From this perspective, smart cities (or smart spaces in general) refer to physical environments in which information and communication technologies and sensor systems disappear as they are integrated into physical objects and the environment in which we live, travel and work (Steventon et al., 2010).
5. Smart cities are defined as territories that bring together innovation systems and ICTs (singular scientific and technical infrastructures) in the same locality, combining the creativity from the talented individuals that makes up the city's population, the institutions that foster innovation and the digital innovation spaces themselves (Intelligent Community Forum [ICF], 2006; Komninos, 2002).

The diversity in the definitions of what smart cities are is due to the multiple scientific and technological along with social movements involved in their creation, i.e., *cybercities*, *smart growth*, *smart communities* and *smart innovation environments*.

5.1.3 Facial Recognition

In recent years, facial recognition has become one of the most studied applications in fields such as biometrics, image processing or pattern recognition (Hernández, 2010).

The main advantage of facial recognition technology is that it allows the analysis and verification of millions of records in matches search, making it possible to automatically and reliably verify a person's identity (Monroe, 2009).

Facial recognition is part of biometric technologies that are systems composed of six subsystems: data collection, data transmission, signal processing, data storage, decision-making, evaluation and performance (Luzmilla et al., 2009). This area is of great importance for scientists and engineers since the 1970s, due to all the efforts made for the proposal the algorithms, and methods are becoming more robust and accurate to introduce the use of this type of systems in everyday life (Santana et al., 2017), because it is one of the few biometric methods that have the advantages of high accuracy and low intrusion.

One of the main reasons for its usage is to encourage the development systems or applications focused on security and surveillance, as they are a powerful tool for management identification. Among these systems, some are capable to learn automatically, acquiring the ability to recognise faces from strangers that constantly are captured by integrated camera.

5.1.3.1 Ethics

Ethics are important when it comes to identifying people through AI. One of the relevant aspects is the right to privacy. The present investigation does not contemplate the intrusion or violation of the right to privacy, and this proposal requires family consent from the family victim to provide data and images that allow the search. It is necessary to clarify that there will be no random or deliberate searches.

UNESCO (United Nations Educational, Scientific and Cultural Organization) is working on the Development of a Recommendation on the Ethics of Artificial Intelligence where the guidelines, regulation, scope and use are established.

5.1.3.2 Typical Issues of Machine Vision

Traditional machine vision systems perform reliably with well-built, consistent parts. They work through step-by-step filtering and rule-based algorithms that are more cost-effective than human inspection. However, algorithms become difficult to develop as exceptions and deficiency libraries increase.

The image quality and texture of complex facial features present a problem for the system when trying to identify a face, thus impairing the accuracy of the individual's analysis.

There are other external aspects that can affect the analysis of the faces, one of them being the lighting present in the place where the individual is, and on the other hand, the case of not being able to identify a face may occur because the person carries some facial accessory (glasses, mask, etc.) (INFAIMON, 2016).

5.1.3.3 Typical Issues of Industry 4.0

The Industry 4.0 development represents a new way of organising production definition. Through it, an important change can be achieved in processes and jobs of the productive factories. However, as in any field, there may be some drawbacks.

Not all organisations are adequately adapting to Industry 4.0. With the constant changes that this implies, many industries are at risk of becoming outdated in a short period of time. Additionally, other companies remain rooted in the traditional system and/or do not dare to make the leap to the new industry, which poses a risk to the company that their systems become obsolete (ALDAKIN, 2017).

5.1.4 CONVOLUTIONAL NEURAL NETWORKS (CNNS)

The heart in the current facial recognition systems is the algorithm. These mathematical algorithms are used within the intelligent system to encode the faces.

Convolutional networks (LeCun, 1989) are a specialised type of neural network that has a known grid-like topology with its functionality based on data processing (Goodfellow et al., 2016).

The name "convolutional neural network" (CNN) indicates that the network employs a mathematical operation called convolution. Convolution is a specialised type of linear operation (Goodfellow et al., 2016). Convolutional networks are simply neural networks that use convolution rather than general matrix multiplication in at least one of their layers.

In many applications, CNNs are now considered the most powerful image classifier and are responsible for driving the state of the art in computer vision subfields that take advantage of machine learning (Rosebrock, 2017).

5.2 PROBLEM

The constant monitoring and control of public places has become a necessity within society. In particular, video camera surveillance systems are useful to detect criminal activity in airports, shopping centers and parking lots, among others. Thanks to their constant development, one of their main objectives is to have automated systems that require little or no human interaction for their operation and detection of potential threats.

According to the figures of the RNPED (National Registry of Lost or Disappeared Persons in Mexico) (SESNSP [Secretariado Ejecutivo del Sistema Nacional de Seguridad Pública], 2018) between 2006 and the cut-off date of April 30, 2018, there was a total of 34,016 records of people whose investigation files corresponded to the description of the location of the disappeared, all of them Mexican.

At national level, few projects have been developed incorporating biometric technologies for the identification of missing people. As a result, a wide variety of algorithms have been proposed and have become a fundamental part of commercial products (Rosas et al., 2012).

Compared to traditional methods, the integration of AI in video surveillance systems improves the certainty and quality of the systems, also scaling up the degree of innovation.

5.3 PROPOSAL

Over the years, security approaches have been developed that help keep confidential data safe, limiting the chances of a security breach.

Facial recognition is considered one of the few biometric methods that have the merits of high accuracy and low intrusion, characterised by using the facial features of an individual within the purpose to automatically identifying and verifying the person from a digital image or a frame from a video source (Hamid et al., 2010; Li, 1993; Rathi et al., 2012).

Machine learning is a very important branch in the AI domain, which involves image recognition, highlighting facial recognition. Besides being one of the most relevant applications of image analysis. Several facial recognition techniques are based on appearance and feature-based approaches (Peace, 2018).

Facial recognition automatically determines whether two faces are likely to correspond to the same person. Face detection is the process of automatically identifying human faces in visual media (Jhanani et al., 2020). A face that is detected is reported on an edge with a matching size and orientation, and at the moment it is detected, reference points such as eyes and nose are searched. Face tracking extends face detection to video sequences. Any face that appears in a video during any period of time is often tracked.

The implementation of a facial recognition system specialised in the identification of missing persons would bring an advantage in the field of security.

In view of the high figures cited in the previous section, the development of an intelligent system was taken as an initiative, due to the seriousness of the situation and the sympathy to the people and families affected.

The present proposal forwards an intelligent model that obtains knowledge by means of a CNN named **AU**tomatic al**G**orithm **US**ed **T**o **U**nify the **S**earch (**AUGUSTUS**). The model is based on the usage of facial recognition for the identification of missing people by means of a system of IP (Internet Protocol) cameras that facilitate obtaining the location point of the subject in real time. At the moment the missing person is identified, an instant alert is sent to the corresponding authorities, helping to intercept the individual and ensuring their well-being.

5.4 METHODOLOGY

The proposed methodology was developed to become directly involved in certain key areas and has seven fundamental moments, which are triggered by a series of actions. These moments are described as follows (left to right in Figure 5.1):

1. The security expert in charge of the system will enter the corresponding information of the individual into the model for further processing.
2. The information is processed, causing unwanted items to be removed.

FIGURE 5.1   Living lab.

3. The information will be classified according to its reference points, e.g. length of jaw line and distance between the eyes with an accuracy of at least 95%.
4. When the system is active and working within the camera network, a search of the faces captured by the model will be started; this allows to compare and identify if there are coincidences within the database.
5. When there is a coincidence with the established percentage, an alert is sent via Gmail from a personalised email corresponding to the model, and this alert is addressed to the corresponding authorities. Within the email, additional information about the name and location of the missing person is attached.
6. Once the email has been sent, the event is registered in the LIIDIA database, with the intention of having a control of the generated information by the model for later queries or the generation of reports.
7. As a final step, the corresponding and nearest authorities must go to the location point sent by the system within the email for the interception of the subject in question (this final responsibility rests solely with the authorities who will resort to perform such action in support of the information provided by LIIDIA).

In facial recognition, people are identified in video frames. That is, the model extracts characteristics of a face in an input image and compares them with the characteristics of the tagged faces within the knowledge base.

The system has the function of mapping the face and creating a faceprint, a unique numeric code for that face. Once LIIDIA has stored a faceprint, it can compare it with the thousands or millions of faceprints stored in its database.

The comparison is based on a feature similarity metric, and the most similar knowledge base input tag is used to tag the input image.

If the similarity value is below a certain threshold, the input image is labeled as unknown.

LIIDIA employs a method to detect a human face within an image. The method includes receiving multispectral image data that include a human face, such as multispectral image data comprising visible light image data and near-infrared (NIR) image data.

The method comprises processing the multispectral image data to detect the human face and may further comprise associating the human face detected in the visible light image data with the face detected in the NIR image data to determine whether a human face is detected.

Comparing two facial images to determine whether they show the same person is known as facial verification (SeungJu et al., 2018).

5.4.1 BIOMETRIC DATABASE

The use of databases within systems capable of identifying people on a large scale (Kak, 2020) plays a vital role, because in them, all the fundamental information is stored, becoming a cog without which the system could not function properly. In addition, the relatively small size of facial profile databases makes facial recognition a very attractive technology.

When we talk about biometric systems, we must remember that there are those called unimodal systems (Faundez-Zanuy et al., 2006), that is to say, they only contain data of one single characteristic or biometric modality, and on the other hand, the systems with databases storing more than one characteristic are called multimodal; the databases used in them can contain information about speech, iris, face, handwritten signature and handwritten text, fingerprints and keystrokes (Fierrez et al., 2010).

LIIDIA is a unimodal biometric system, because it is focused on facial recognition. Therefore, it is important to mention that two types of databases are used. In the first one, the photographs provided by the families are stored; these photographs are the ones that directly feed the intelligent system; the second database is in charge of storing the person's private data and the registry as soon as it is detected by the camera network.

Because sensitive data information is handled, it is important to mention that these databases are protected, and the data obtained from the storage is not shared with any public or private organisation. It is also important to note that the input data is provided by the families of the victims giving their previous consent to add this information to the knowledge base of the system.

5.4.2 CAMERA NETWORK ARCHITECTURE

The video frames are extracted for analysis from the video processor in which the live transmissions from the cameras are stored. Many factors can influence the images processing, and it is important to consider that a human face is not only a 3D object, it is also a nonrigid body. In addition, facial images are often taken in a natural environment. That is, the background of the image may be complex, and the lighting condition may be drastic (Li, 1993). Therefore, a reliable video extraction system is necessary.

A bridge between the camera network and the face recognition system was implemented using the Real-Time Streaming Protocol (RTSP). The cameras used in the implementation of the project include a 360° panning camera and several "Bullet" ones, all of them monitored with a HikVision Network Video Recorder (NVR).

Both the panning camera and the bullet ones have a maximum resolution of 2 megapixels; they have a progressive scan CMOS (complementary metal oxide semiconductor) image sensor and are capable of detecting motion in specific points of interest. The main difference between the two models is the structure and protection they have, which gives them different capabilities, and the panning camera (see Figure 5.2) has a protection "dome" around the lens, which provides it with weather protection and makes it harder to be vandalised. The dome structure gives these cameras a wide range of view and the possibility of adding infrared light emitting diodes (LED), which allow them to work even in low light conditions; however, this structure is also the one preventing them from having long range, and the interest points of these cameras must be in a close range to maintain a reliable resolution. Bullet cameras, on the other hand, have the "classic" camera structure, an elongated rectangular housing with almost no protection, but the simplicity of their design allows them to have wider lenses, which also provides long-range camera resolution. Both cameras, working together in a well-designed network, provide an optimised monitoring environment, perfect for the needs of this project.

The whole network is connected to the NVR DS-7732NI-I4/16P, which can be operated both ways: local and remote. It can support up to 4 HDD of 6 TB each and a maximum of 32, 2 megapixels, channels. It is worth mentioning that the NVR by itself is capable to detect movement in multiple channels at the same time, which makes it convenient for different applications where the user wants to store footage of only specific moments or locate places where a person is moving at the moment. Finally, the most important aspect of the camera network is the capability to communicate with other devices via RTSP (see Figure 5.3), which makes it easier for the project due to the standardisation of the communication commands, allowing it to stream video signals by only selecting in the NVR the desired channel.

The RTSP is a level protocol application for control over the delivery of real-time data such as audio and video; it establishes and controls one or more time-synchronised broadcasts of continuous media (Schulzrinne et al., 1998). Through RTSP, the algorithm identifies each camera with the information provided by the NVR, using their IP as an identifier, to know the exact location of each one, selects the channel to analyse, and the video is transmitted to the website where, from the backend, it

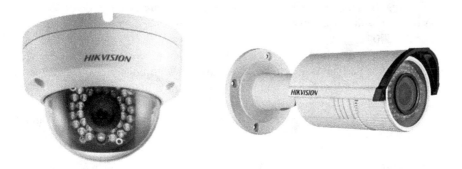

FIGURE 5.2 HikVision panning camera model DS-2CD2120F (left), HikVision bullet camera model DS-2CD2622FWD (right).

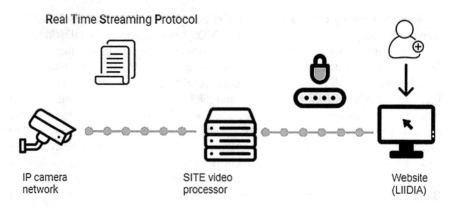

Real Time Streaming Protocol

IP camera
network

SITE video
processor

Website
(LIIDIA)

FIGURE 5.3 Sequence of actions to obtain the video in real time.

is first transformed into photo frames. After that, they are consumed by the model that is in charge, the frames are analysed in search of a detected face and, if so, the comparison is made in the knowledge base; if a similarity is found, the queries are executed to extract the personal information of the identified individual, thus activating the actions of the notification module.

5.4.3 WebApp Design

Currently web applications have become complex systems, with interfaces seeking to be intuitive for users. Therefore, improving architecture and design techniques have become a primary task with the intention of offering a better user experience.

The user interacts with these web applications through a browser and consequently requests are sent to the server, where the application is hosted and which normally makes use of a database in which all the information related to the application is itself stored. The server is responsible to process the information and returning a response to the user. In summary, the system is made up of: the browser, with the function of rendering the user interfaces, and the application, which is in charge of performing the operations according to the actions that are executed in the interfaces and interact with the database. This distribution is known as layered architecture (Garrido, 2004).

LIIDIA has a dynamic web interface, the function of which is to allow the visualisation of the various video channels that are live within the network of cameras, offering the possibility of interleaving between them, with the intention of allowing the user to monitor from your perspective.

For the web application construction, the following technologies and programming languages were used:

1. *HTML* (HyperText Markup Language): This is a markup language that defines the structure of the content, and it is represented with a series of elements called tags with the intention of adjusting the parts of the content.

2. *CSS* (Cascade Style Sheet): This technology is used to describe how the HTML elements should be displayed on the screen; it allows you to control the design of web pages.
3. *Flask:* This is a microframework developed in the python programming language that offers simplicity and power, provides tools, libraries and technologies that allow you to create a web application and avoid repetitive code.
4. *Bootstrap:* This is a CSS framework for the design of web applications that allow to create user interfaces offering a better experience for browsing.

5.4.4 NOTIFICATION MODULE

The proposed model has a notification module via Gmail that is activated instantly. It finds in its knowledge base a similarity of 95% to some face captured by the security cameras vision. When this happens, the system moves to perform a series of queries in the database to extract the data of the individual and runs a search in a data dictionary that contains the exact location of the installation of the camera that captured the face, performs the search according to the IP assigned to the camera and returns the string with the location (see Figure 5.4).

Once we have the corresponding data, we proceed to move them to a special class where the email is built to be sent to the authorities. The data is attached to the email in such a way that it can be understood by the user who receives the email.

The sending of personalised mail is achieved through the SMTP (Simple Mail Transfer Protocol). This protocol is used as the common mechanism for transporting email between different hosts within the Transmission Control Protocol/IP (TCP/IP) package. In SMTP, an SMTP client process opens a TCP connection to a server SMTP process on a remote host and attempts to send mail over the connection (Riabov, 2007).

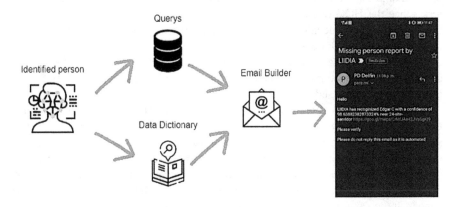

FIGURE 5.4   Sequence of actions in the notifications module.

5.4.5   AUGUSTUS (CNN)

AUGUSTUS is a search algorithm (in this case, for missing people), which consists of three dependent modules that, together, are capable to identify the facial features of each person. These modules are:

1. Facial recognition and alignment
2. Obtaining characteristics
3. Classification of characteristics

Derived from the nature of the problem it intends to solve, it has conditions for the project, such as: few training images; because when a person is reported missing, the family only provides between one and four photos of the missing person. Also, the positioning of the cameras means that people are rarely looking directly into the lens. Based on these characteristics, a way was sought to make the algorithm more precise, and it was decided to divide it into the three main modules.

This algorithm is based on the OpenFace implementation idea, which it is a deep learning facial recognition model developed by Brandon Amos, Bartosz Ludwiczuk and Mahadev Satyanarayanan. It is based on the article: "FaceNet: A Unified Embedding for Face Recognition and Clustering" by Florian Schroff, Dmitry Kalenichenko and James Philbin at Google.

5.4.5.1   Facial Recognition and Alignment

Due to the problems caused by the elevated position of the cameras, the first process is facing recognition and their alignment. In itself, the algorithm is designed to work with video in real time, so it would be unnecessary processing to pass each frame through each stage of the algorithm; thus, facial recognition also serves as a filter to decide which frames pass through the algorithm and results.

For face recognition, the histogram of oriented gradients (HOG) method is used to find particular characteristics of human faces; this is done using a sliding window of fixed size that runs through the entire image and calculates the HOG descriptors. Once the HOG descriptors are found, they are classified with support vector machines (SVMs) to decide if they are characteristics of a human face or not. Finally, the windows, the HOG descriptors of which have been classified as characteristics of human faces are taken and combined to find the bounding frames of each face contained in the video frame. Because of this, the implementation of this method of the "dlib" library is used.

In the case of face alignment, once the bounding box is in place, a shape predictor is used, implemented in the same "dlib" library, which the position of important facial landmarks such as the corners of the eyes and nose are obtained. Subsequently, the geometric transformation of images is used, precisely the affine transformation of OpenCV, where the positions of the facial reference points obtained, and the standard positions defined are used so that these reference points (eyes and nose) always appear in the same location in each frame. At the end of the process, the faces are resized to the same size of 96×96 pixels.

5.4.5.2   Obtaining Characteristics

Regardless of the characteristics obtained with the HOG descriptors, since it is necessary to have characteristics that identify each human face as unique and because of the peculiarity of having few training images, the usage of a CNN is proposed, which contains specific parameters for each of the neurons, and this is achieved through knowledge transfer of a CNN previously trained with a database of many more images. This saves a lot of training time, as well as substantially improving the accuracy of the system even with few images.

For this specific case, the OpenFace CNN is used with 3,733,968 parameters obtained from the "nn4.small2" model, the training of which was done with a combination of the two largest facial recognition data sets (until August, 2015), publicly available with the names: FaceScrub and CASIA-WebFace. Kera's OpenFace implementation is done.

One of the main features of this OpenFace implementation is the existence of a fully connected layer with 128 hidden units followed by an L2 normalisation layer on top of the convolutional base. These top two layers are called the embedding layer from which the 128-dimensional embedding vectors can be obtained. After the image with the face is sent through the trained neural network in a single forward pass, these 128 facial embedding are obtained, and by means of a grouping algorithm, classification and detection of similarities are achieved.

5.4.5.3   Characteristics Classification

As a third layer in the implementation of the algorithm, SVMs independent of the SVMs used in the first stage of the algorithm for face recognition in general were used; at this point, SVMs are used for the clusters creation from the characteristics found by the neural network from each of the faces, this to better classify each face since the SVM continues to be effective even in occasions in which the number of dimensions is greater than the number of samples as is common in the case of images of missing persons.

SVM works on a dimensions plane with data that is to be classified; for this, it makes use of a hyperplane which serves as a border to delimit each of the clusters formed of data of each class. Since SVM is an algorithm with supervised training, it is necessary to train with labeled data to later make the hyperplane that best delimits even if it is not linear.

The implementation of the SVM algorithm made by the Sklearn library was used, which greatly facilitated the training of this algorithm and its subsequent deployment to classify faces.

5.4.5.4   AUGUSTUS Flow

The algorithm begins with obtaining the video information transmitted by the camera. Once the video is obtained, we proceed to capture each frame and search for faces in it, and if the face is not obtained, the algorithm stops to process the next frame. In case a face has been found, this face is cut out and digital processing is applied to try to align it as much as possible, to improve the quality of the following step (see Figure 5.5).

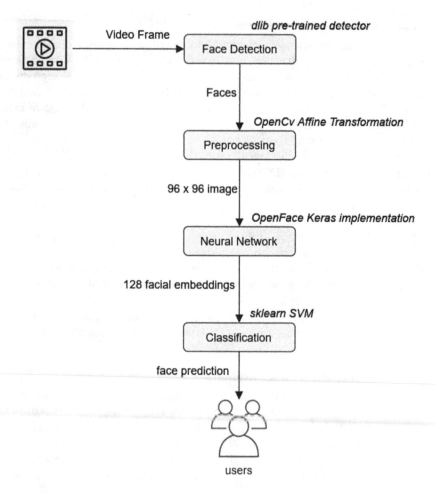

FIGURE 5.5   Face recognition algorithm flow.

Once the face is aligned, it passes through the CNN, which returns a single numerical value depending on the specific characteristics of each face. This numerical value is used to feed the SVM, which creates clusters with the characteristics of each face obtained from each person image, to later classify them and, thus, be able to predict which face corresponds to each person.

5.4.5.5   Infrastructure

The power process is required when working with AI exceeds that offered by conventional desktop computers; a computer with specific characteristics is necessary, that is why the present proposal is housed in a high-performance server with characteristics such as: Xeon Bronze 3106 8C 1.7-GHz rack processor with mounting rails, disk controller card with RAID 5.6 and ten 12-GB SAS/SATA, 64 GB RDIMM RAM, four disks of 8TB SAS 7.2K HotSwap and dual power supply. This server concentrates the computing power that has been contributed voluntarily by different teams to store, but also to process, the tasks described in this proposal.

Currently the average computational power of the server is 42.54 GFLOPS (floating point operations per second); one GFLOPS is equal to 10^9; therefore, the server is capable of processing 42,540,000,000 operations per second (real, GHz-independent operations).

5.5   RESULTS

This report section shows the preliminary results related to the implementation model within a living lab, after its testing.

Several tests were performed to ensure that this model will function properly, which directly influenced the system's precision adjustment around the capture of face characteristics by the intelligent model, significantly reducing false positives.

Figure 5.6 shows the results of these tests; one of them was to ensure that the model will indicate when a face is detected but, at the same time, is not within its knowledge base, another important test was to match two test subjects to ensure that the model is able to recognise both faces that are within the field of camera vision.

5.6   CONCLUSION

Mexico is a great country that has the potential and need to development the technical side, mostly focused on security programs, and the main proposal for this development is to create and offer tools that will be able to identify missing people, because these is a social problem that cause a great impact.

This software was born with a single purpose and that is to join the cause in the search for a missing person. As a conclusion of the executed tests, the development of the project is defended because this AI would evidently turn out to be a quite effective tool tested in the network of security cameras in streets, airports and centers where there is a high concentration of people where it is evidently more difficult to identify a person's face; with the support of this AI, it is possible to take advantage of

FIGURE 5.6 (Left) Face labeled "unknown", due to the similarity being below the imposed threshold. (Right) Recognition of persons registered in the knowledge base as missing.

this situation and, according to the situation, reunite the victims with their families with the help of the intervention of the corresponding authorities.

It is important to start to develop comprehensive and coordinated approaches that use new technologies alongside known surveillance techniques and collection, distribution and management techniques to maximise the value using biometric recognition technologies.

REFERENCES

ALDAKIN, S.F., 2017, Industria 4.0. http://www.aldakin.com/industria-4-0-que-es-ventajas-e-inconvenientes/.

California Institute for Smart Communities (CISC), 2001, "Ten Steps to Becoming a Smart Community". Retrieved from http://www.smartcommunities.org/library_10steps.htm.

M. Chowdhury, A.W. Sadek, 2012, "Advantages and Limitations of Artificial Intelligence." Artificial Intelligence Applications to Critical Transportation Issues, vol. 6, no. 3, 360–375.

P. Droege, ed., 2008, "Intelligent Environments: Spatial Aspects of the Information Revolution", Elsevier, Faculty of Architecture Urban Design Program, The University of Sydney, Sydney, Austral.

M. Faundez-Zanuy, J. Fierrez-Aguilar, J. Ortega-García, J. Gonzalez-Rodriguez, 2006, "Multimodal Biometric Databases: An Overview". In IEEE Aerospace and Electronic Systems Magazine, vol. 21, no. 8, pp. 29–37, Aug. DOI: 10.1109/MAES.2006.1703234. [Mismatch]

J. Fierrez, J. Galbally, J. Ortega-Garcia, M.R. Freire, F.A.F.D. Ramos, D.T. Toledano, et al., 2010, "BiosecurID: A Multimodal Biometric Database." Pattern Analysis and Applications, vol. 13, no. 2, 235–246.

J.C. Garrido, 2004, "Arquitectura y diseño de sistemas web modernos". *InforMAS*, Revista de Ingeniería Informática del CIIRM, vol. 1, 1–6.

I. Goodfellow, Y. Bengio, A. Courville, 2016, "Deep Learning", MIT Press. Available in http://www.deeplearningbook.org.

R.A. Hamid, J.A. Thom, 2010, "Criteria That Have an Effect on Users While Making Image Relevance Judgements". In Proceedings of the Fifteenth Australasian Document Computing Symposium, pp. 76–83.

R.G. Hernández, 2010, "Estudio de Técnicas de Reconocimiento Facial". Escuela de Ingeniería de Telecomunicaciones de Barcelona, Universidad Politécnica de Catalunya, España.

INFAIMON, 2016, "La visión artificial en la industria 4.0". https://blog.infaimon.com/la-vision-artificial-industria-4-0/.

Intelligent Community Forum (ICF), 2006, "What Is an Intelligent Community?" Retrieved from http://www.intelligentcommunity.org/displaycommon.cfm?an=1&subarticlenbr=18.

R. Jhanani, S. Harshitha, T. Kalaichelvi, V. Subedha, 2020, "Mobile Application for Human Facial Recognition to Identify Criminals and Missing People Using TensorFlow". *International* Journal of Research in Engineering, Science and Management, vol. 3, no. 4, 16–20.

Amba Kak, ed., 2020, "Regulating Biometrics: Global Approaches and Urgent Questions", AI Now Institute. https://ainowinstitute.org/regulatingbiometrics.html.

N. Komninos, 2002, "Intelligent Cities: Innovation, Knowledge Systems and Digital Spaces", London and New York, Spon Press.

Y. LeCun, 1989, "Generalization and Network Design Strategies. Technical Report". CRG-TR-89-4, University of Toronto, vol. 19, 143–155.

S.H. Li, 1993, "An Introduction to Face Recognition Technology". *Informing Science Special Issues on Multimedia Information Technologies*, vol. 3, 1–7.

Mg. Luzmilla, Jc. Gonzáles, W. Contreras, C. Yañez, 2009, "Tecnologías Biométricas Aplicadas a la Seguridad en las Organizaciones". Facultad de Ingeniería en Sistemas e Informática, Universidad Nacional Mayor de San Marcos, Revista de investigación de Sistemas e Informática, vol. 6, no. 2, 55–56.

P. Mayumbo. 2018. "An Investigation on the Use of LBPH Algorithm for Face Recognition to Find Missing People in Zimbabwe". International Journal of Engineering Research & Technology (IJERT), vol. 7, (07), 80–86.

D.A. Monroe, 2009, "Method for Incorporating Facial Recognition Technology in a Multimedia Surveillance System." U.S. Patent 7,634,662, issued December 15.

M. Peace, 2018, "An Investigation on the Use of LBPH Algorithm for Face Recognition to Find Missing People in Zimbabwe". International Journal of Engineering Research & Technology (IJERT), vol. 7, no. 7, 80–86.

R. Rathi, M. Choudhary, B. Chandra, 2012, "An Application of Face Recognition System using Image Processing and Neural Networks". International Journal Computer Technology Application, vol. 3, no. 1, 45–49.

V.V. Riabov, 2007, "SMTP: Simple Mail Transfer Protocol". The Handbook of Computer Networks, Vol. 2, edited by H. Bidgoli, John Wiley & Sons Publishing Company, New York, NY, September 2007.

U.O. Rosas, M.S.G. Vázquez, A.A.R. Acosta, 2012, "Algoritmos de procesamiento del iris para un sistema de reconocimiento biométrico". Research in Computing Science, vol. 55, 353–362.

A. Rosebrock, 2017, "Deep Learning for Computer Vision with Python". PyImageSearch, vol. 1, 23–25.

M.A.G. Santana, L.E.D. Sánchez, I.T. Paz, M.H. Romero, 2017, "Estado del arte en reconocimiento facial". Research in Computing Science, vol. 140, 19–27.

H. Schulzrinne, A. Rao, R. Lanphier, 1998, RFC2326: Real Time Streaming Protocol (RTSP).

SESNSP (Secretariado Ejecutivo del Sistema Nacional de Seguridad Pública), 2018, Gobierno de México. Registro Nacional de Datos de Personas Extraviadas o Desaparecidas, RNPED. https://www.gob.mx/sesnsp/acciones-y-programas/registro-nacional-de-datos-de-personas-extraviadas-o-desaparecidas-rnped (cited 2021 Jan 14).

H.A.N. SeungJu, M. Ko, K.I.M. Deoksang, H.A.N. Jae-Joon, 2018, U.S. Patent Application No. 15/837,170.

A. Steventon, S. Wright, eds., 2010, "Intelligent Spaces: The Application of Pervasive ICT", Springer Science & Business Media, London.

6 Inclusion of Impaired People in Industry 4.0

An Approach to Recognise Orders of Deaf-Mute Supervisors through an Intelligent Sign Language Recognition System

Martín Montes Rivera, Alberto Ochoa Zezzatti and Luis Eduardo de Lira Hernández

CONTENTS

DOI: 10.1201/9781003122401-6

6.1 INTRODUCTION

Industry 4.0 is the term for referring to the fourth industrial revolution, counting from the first mechanical loom in 1784 (Industry 1.0); the use of electricity and mass production in 1870 (Industry 2.0); the automated processes with programable logic controllers and information technology systems in 1960 (Industry 3.0) [1–3]. This index for the industrial revolutions was introduced in 2011 during the Hannover Fair to support and address future challenges in industrial environments [1].

The Industry 4.0 technologies include machine-to-machine (M2M) communications, Internet of Things (IoT), cyber-physical systems (CPSs), Big Data analytics (BDA), robots, cobots, artificial intelligence (AI), virtual reality (VR), augmented reality (AR), among others [1, 4].

Adopting the Industry 4.0 technologies brings advantages to the supply chain, including integrating technologies, better connectivity levels, comprehensive integration, performance improvements, flexibility, among others [4]. On the other hand, using these technologies demands more specialised workers and robotisation with AI has increased unemployment [1, 4].

Nevertheless, new technologies such as AI, robotics, IoT and AR, included in Industry 4.0, have shown in other scenarios that they boost human capabilities while raising productivity [5].

Boosting communications for suppressing barriers is now possible in Industry 4.0 through inclusion environments with IoT, AI, AR, especially for those who suffer from conditions that limit their interactions in conversations, sounds, commands and warnings crucial in industrial environments [6].

Sign language is a tool for simplifying interactions for 5% of the world's population suffering the deaf-mute condition. However, industrial environments have several noises and situations that affect this communication channel threatening security. This situation makes workers with hearing impairment or deaf-mutes struggle for job positions [6–8].

Several researchers have built sign language recognition systems classified into device-based and device-free options. Device-free sign recognition systems have lower accuracy than device-based solutions, but they are less expensive than device-based systems and do not require wearable devices [9–11].

Examples in the literature for device-based sign recognition systems implemented include a proposal [12], which allows controlling an arm kinematic robot using the prebuild Leap Motion sensor to detect the sign gestures. However, it requires plugging it into a personal computer with a limited distance for detecting the instructions.

One group [8] proposes other device-based sign recognition systems with a wearable device used in hand with accelerometers interpreted by a Raspberry PI with a support vector classifier that identifies the signs expressed in an LCD screen. However, the system is unsafe for industrial activities because the wearable could become trapped in the manufacturing process's machines.

Similarly, another study [13] uses a glove with hand accelerometers interpreted by a Raspberry PI to interpret signs for deaf-mute people, and then it uses Amazon voice services to produce voice outputs. This approach improves mobility, but the glove still being dangerous when interacting with engines and moving machines.

Another novel device is used backward [13], because it translates voices to deaf-mute sign language. This device records audio voices and transforms them into text. After that, use the text for changing the works to familiar signs in the deaf-mute language.

One proposal [14] is a device-based system for recognising hand signs using convolutional neural networks (CNNs) for classifying electromyography signals associated with 12 different hand signs reaching an accuracy of 98% of accuracy. However, the wearable device could still affect security in Industry 4.0.

External devices also include cell phones [14], where an application uses a keyboard designed to allow deaf-mute people to communicate with others producing audio responses [14]. The user either works with a text-keyboard or a signs-keyboard to produce the sound outputs. However, working in Industry 4.0, paying attention to a cell phone also threatens security.

On the other hand, device-free signs recognition systems use an image-based approach that requires robustness against the light, color and dynamic variations, which usually demand high computational resources in image classification and object recognition tasks [9].

CNNs and deep learning improved image classification and object recognition accuracy compared with the image processing techniques. Moreover, new architectures of CNNs are available for devices with low processing capabilities such as the MobileNet, ShuffleNet, SqueezeeNet, among others [15].

A device-free sign recognition system using binocular vision has also been proposed [11]. The system benefits from the 3D perception of the binocular vision for isolating the background's hands. However, the final results are not conclusive, exhibiting low accuracy and poor processing speeds related to working with information perceived through two cameras.

Similarly, a hand gesture recognition system that works with 3D perception based on an RGB-D (red, green, blue-depth camera) camera that subtracts the background and focuses on the hand's position has been proposed [16]. The hands detected then are processed with a CNN that identifies the fingers position with up to 97.12% accuracy in their 3DCNN proposed model that considers time and space perception.

Alternatively, monocular vision stills being used with CNNs, as shown in Ref. [17], where they identify hand signs in infrared images perceiving the hand's pointing position in the board to map it on the screen for online students. The study reports 92.26% efficiency estimating the ploting position.

In this work, we propose an inclusive environment for deaf-mute people in Industry 4.0 with a sign recognition system based on MobileNet2 CNN proposed by Google, using a simple RGB camera. We use the game dataset in Ref. [18] generated by Laurence Moroney, who leads the AI Advocacy at Google. The possible commands associated with the three detected signals in that game include "Emergency Stop," "Warning Pause," and "Run Process," which are standard instructions in assembly lines of manufacturing processes.

Our proposal improves the results in other works because it detects the hands and signs correctly without 3D information or particular infrared images. We achieve this by improving the dataset's representativeness augmenting it with variable conditions of light, colors, rotations, translations and our proposed function for randomly changing the background. The proposed data augmentation covers situations that could occur in an industrial environment.

Our solution allows deaf-mute supervisors to transmit orders in noisy industrial environments. The system proposed detects three possible signals well-identified in the famous game paper, rock and scissors.

6.2 THEORETICAL FRAMEWORK

6.2.1 Digital Images

6.2.1.1 Color Representations
They are the digital representation of colors in digital images through different models enhanced for color segmentation, light robustness, device proprieties, among other applications [19].

6.2.1.1.1 RGB Color Model
RGB represents the three additive primary colors, which are merged as one color to create new colors of the visible spectrum, as in Figure 6.1 [19].

The cube normalised RGB representation has gray values on the central yellow diagonal going from (0,0,0) or the black color to the white color on the opposite

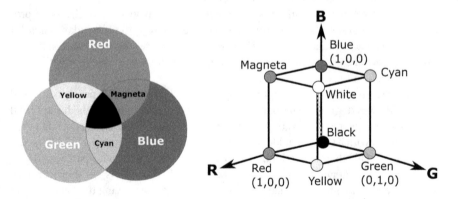

FIGURE 6.1   RGB color model in primary colors and color cube representation.

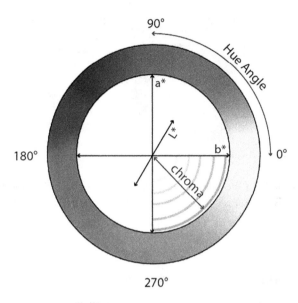

FIGURE 6.2 CIE L*a*b* color model [21].

corner (1,1,1). Most image applications use the RGB space as the base color model since the acquired images do not require any further transformation to show them on the screens [19, 20].

6.2.1.1.2 CIE L*a*b

The second uniform color space originating from the CIE XYZ space in 1976, with a white reference point, is often called the CIE L*a *b color model. The color model L*a*b* specifies the color based on its position in a 3D color space. The components of L* are the lightness of the color (when L* = 0 means black and when L* = 100 means while) and the chroma* (red for positive values and green for negative values) and the hue b* (yellow for positive values and blue for negative ones) as illustrated in Figure 6.2. CIELAB is device-independent [20, 21].

6.2.1.1.3 HSV

The HSV (hue, saturation, value) uses a conical representation of RGB points (Figure 6.3). The H and S parameters used in color segmentation are the chrominance components, and they change the colors in the same way that humans perceive the color spectrum. The V component is segregated from the chrominance components and represents the quantity of light [19, 20].

6.2.1.2 Chromatic Distance or Color Difference

The chromatic distance measures the difference between two colors. The most common formula for measuring it is the Euclidean distance. However, the human eye has a different sensitivity to the individual component of RGB space. Thus, Euclidean distance is not a precise approach, but this is not an issue in deciding whether a

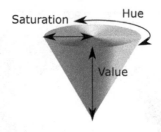

FIGURE 6.3   HSV conical spectrum space.

sample corresponds to the white color. Equation (6.1) describes the Euclidean distance in the RGB space for calculating the chromatic difference [22].

$$\Delta E_{RGB} = \sqrt{\Delta R^2 + \Delta G^2 + \Delta B^2} \qquad (6.1)$$

6.2.2   CONVOLUTIONAL NEURAL NETWORKS

CNNs are a specialised type of artificial neural networks with a grid-like topology that applies an operation called convolution [23].

CNNs consist of three steps (see Figure 6.4). The first layer or detector layer executes multiple convolutions combined with linear activations. Then, nonlinear activation functions run through each linear activation. After that, a pooling layer or subsampling layer allows the sheet's output to change for another detector level or processes the final output in the prediction layer [23].

6.2.2.1   Pooling Layers

A pooling function substitutes a summary statistic of the surrounding outputs for the net's output at a particular location. The average of a rectangular neighborhood, the L^2 standard of a rectangular neighborhood, or the maximum value are typical pooling functions [23].

Pooling aims to make the representation roughly invariant to minor translations of the input in all situations. Invariance of translation means that any of the pooled outputs' values do not change if we translate the input by a tiny number. The usage of pooling makes the network invariant to minor translations, and the statistical reliability of the network can be significantly improved [23].

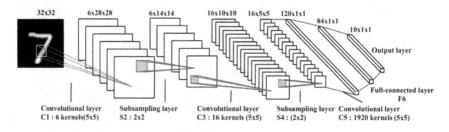

FIGURE 6.4   LeNet-5 convolutional neural network [24].

Pooling also allows the managing of different size images. The general approach is adjusting the size of an offset between pooling regions so that, irrespective of the input size, the classification layer still provides the same output [23].

6.2.2.2   Prediction Layer SoftMax

The SoftMax function defined in Equation (6.2) is the expansion of more than two values of the sigmoid function; it identifies the correct class in ranges [0,1], with n the number of outputs [23, 25].

$$\sigma(x)_i = \frac{\exp(x_i)}{\sum_{j=1}^{n} \exp(x_j)} \tag{6.2}$$

6.2.2.3   Adam Algorithm

Stochastic gradient optimisation is a common alternative in machine learning problems since they require minimising or maximising cost functions. Adam is a gradient descent algorithm specially used when training deep neural networks, which only requires determining the first order gradients and low memory consumption [26].

The Adam algorithm starts by receiving a noisy cost function $f(\theta)$ differentiable concerning its θ trainable parameters. The Adam algorithm optimises the cost function by observing the function's expectation $\mathbb{E}[f(\theta)]$ for random stochastic observations (mini-batches) at time steps $f_1(\theta), ..., f_T(\theta)$ for $1,...,T$ samples [26].

$g_t = \nabla_\theta f_t(\theta)$, the first-order gradient of the cost function at the t sample allows calculating the moving averages of the gradient m_t and the square gradient v_t based on the moving averages $\beta_1, \beta_2 \in [0,1)$, as shown in Equations (6.3) and (6.4).

$$m_t = \beta_1 \cdot m_{t-1} + (1-\beta_1) \cdot g_t \tag{6.3}$$

$$v_t = \beta_2 \cdot v_{t-1} + (1-\beta_2) \cdot g_t^2 \tag{6.4}$$

After that, Equations (6.5) and (6.6) determine the moving averages or momentums with bias correction for obtaining the correct update directions for the trained parameters.

$$\hat{m}_t = \frac{m_t}{\left(1-\beta_1^t\right)} \tag{6.5}$$

$$\hat{v}_t = \frac{v_t}{\left(1-\beta_2^t\right)} \tag{6.6}$$

Finally, the updating rule in Equation (6.7), with the learning rate α and ∂ a correction value that avoids division by zero, allows training the θ parameters.

$$\theta_t = \theta_{t-1} - \alpha \cdot \hat{m}_t / \left(\sqrt{\hat{v}_t} + \partial\right) \tag{6.7}$$

Algorithm 6.1 describes the Adam optimiser general implementation described in Ref. [26].

Algorithm 6.1: Adam Optimiser [26]

Input: $\alpha, \beta_1, \beta_2, f(\theta), \theta_0$

$$m_0 = 0$$
$$v_0 = 0$$
$$t = 0$$

while θ_t not converged do:

$$t = t+1$$
$$g_t = \nabla_\theta f_t(\theta_{t-1})$$
$$m_t = \beta_1 \cdot m_{t-1} + (1-\beta_1) \cdot g_t$$
$$v_t = \beta_2 \cdot v_{t-1} + (1-\beta_2) \cdot g_t^2$$
$$\hat{m}_t = m_t / (1-\beta_1^t)$$
$$\hat{v}_t = v_t / (1-\beta_2^t)$$
$$\theta_t = \theta_{t-1} - \alpha \cdot \hat{m}_t / (\sqrt{\hat{v}_t} + \delta)$$

end while

Output: θ_t ■

6.2.2.4   TensorFlow

TensorFlow is a machine learning method that works in heterogeneous environments and on a wide scale. TensorFlow uses dataflow graphs to represent the shared state and its operations. It maps data flow graph nodes over several computers in a cluster and across many computing devices within a machine, including multicore CPUs, general-purpose GPUs and custom ASICs known as tensor processing units (TPUs). TensorFlow allows developers to experiment with new optimisations and training algorithms. TensorFlow serves several applications, emphasising deep neural networks training and inference. In development, many Google platforms use TensorFlow [27]. The diagram in Figure 6.5 shows the general schema followed in TensorFlow for training an input dataset.

6.2.3   DATA AUGMENTATION

Easy transformations such as horizontal tossing, color space augmentations and random cropping are the first demonstrations demonstrating the usefulness of data augmentations. Such improvements encode many of the previously mentioned invariances that present challenges to image recognition tasks [28].

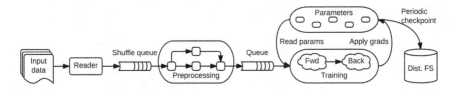

FIGURE 6.5   General schema of TensorFlow for training an input dataset [27].

6.2.3.1 Bright

Photometric methods change the RGB channels by shifting each pixel to different new channel values according to predefined desirable properties. Bright transformation is a standard photometric method used to modify the amount of light in an image in data augmentation, allowing improved robustness against light changes. The augmentation of datasets with light changes implies transforming RGB space images to the HSV representation [28].

6.2.3.2 Rotation

Transformations for the picture's angular movement on the right or left on an axis between 1° and 359°. Slight rotations such as 1–20 or 1–20 can benefit digital recognition activities, making classifiers robust against minor changes in their orientation [28].

6.2.3.3 Translation

Transformation operation for moving images left, right, up, or down transformation improves classifier robustness against small translations. For instance, if all the images in a dataset are in the center, the model demand images perfectly centered. The translation process fills the residual space with values such as 0 or 255 or random or Gaussian noise. This padding retains the spatial dimensions of the post-augmentation picture [28].

6.2.3.4 Share

Shear action changes the image structure by spinning it around a particular axis, helping the vision enhance robustness further with changes in the perceiving camera's tilt; this process multiplies a matrix to shift each pixel's position based on the desired tilt [29].

6.2.3.5 Flip

Flipping is a strategy for data augmentation that copies the input image on a vertical or horizontal axis by multiplying it by a transformation matrix that rotates all the image points around the center of the image, retaining columns and rows, depending on the axis of flipping [28, 30].

6.2.3.6 Zooming or Scaling

Transformation of scaling adjusts the image size through multiplying by a transformation matrix that modifies the image width and height, enhancing robustness against size changes as defined in Ref. [29].

6.3 METHODS

Our proposal starts reading the original dataset and expanding it with the augmentation operations described in Section 6.2.3. Then, we applied our data augmentation operation for changing the background. After that, we train the MobileNet2 CNN with the Adam algorithm. Then we test the accuracy. Finally, we send the classification model to our application for classifying the acquired images and transmitting the corresponding process instructions. The flowchart in Figure 6.6 describes our methodology for developing the proposed deaf-mute assistant.

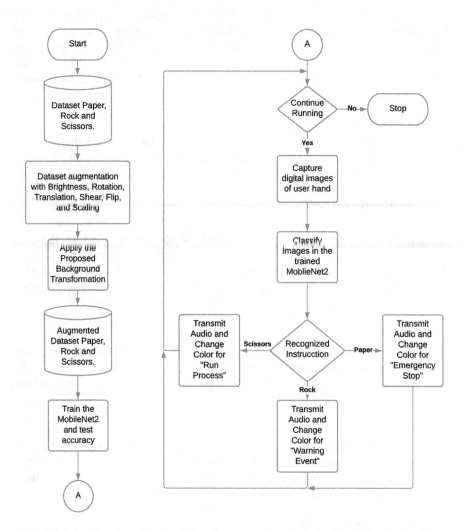

FIGURE 6.6 Flowchart for describing the proposed method for developing the deaf-mute assistant.

6.3.1 RANDOM BACKGROUND CHANGING FUNCTION

Let I_m (height, width, color) be an image of the dataset, with colors defined in terms of the $\mathbb{R}^3$ RGB space containing the RGB tones.

The detection of regions for background changing begins by detecting the current white background color in the dataset images, which we replace with the proposed generated background.

Since the white color is the maximum value perceived in the digital representation, we must identify the maximum values per color with $r_{max} = \max[I_m(\text{height},\text{width},1)]$, $g_{max} = \max[I_m(\text{height},\text{width},2)]$, and $b_{max} = \max[I_m(\text{height},\text{width},3)]$.

We then determine d_c the chromatic distance with Equation (6.8), which is the difference between the red $r_{ij} = I_m(i,j,1)$, the green $g_{ij} = I_m(i,j,2)$ and the blue $b_{ij} = I_m(i,j,3)$ tones of a single-pixel located in the i row and j column of I_m. We use this approach instead of Euclidean distance for reducing computer calculations.

$$d_c = \frac{(r_{ij} - g_{ij}) + (b_{ij} - g_{ij}) + (b_{ij} - r_{ij})}{3} = \frac{2b_{ij} - 2g_{ij}}{3} \qquad (6.8)$$

We also determine d_l the distance between the maximum values corresponding to white and the current examined pixel, replacing Euclidean distance for reducing computer calculations with Equation (6.9).

$$d_l = \frac{(r_{max} - r_{ij}) + (g_{max} - g_{ij}) + (b_{max} - b_{ij})}{3} \qquad (6.9)$$

After that, we obtain $R_1 = [0, n_b]$ a random parameter used for determining the k index for selecting the background B_{ijk} pixel for transformation, where n_b denotes the available backgrounds, i is the row and j is the column of the background. $R_2 = [0,1]$ a random parameter for maintaining the probability of transformation p_t. $R_3 = [1,2,3]$ a random mode for let the color values in the image without change, change the red by blue values, or change the red by green values, respectively.

The R_1, R_2 and R_3 parameters make that the background variability improves robustness because it makes possible hand detection independent from the used background.

Finally, using the previous definitions, we change the background with the selected random modes and background, based on the p_t probability following the steps in the proposed Algorithm 6.2.

Algorithm 6.2: Changing Background Transformation

Input: I_m, p_m

$$\text{height, width, color} = \text{Shape}(I_m)$$

$$r_{max} = \max[I_m(\text{height},\text{width},1)]$$

$$g_{max} = \max[I_m(\text{height},\text{width},2)]$$

$$b_{max} = \max[I_m(\text{height},\text{width},3)]$$

$$R_1 = [0, n_b]$$

$$R_2 = [0,1]$$

if $R_2 \le p_m$ then:
for $i = 1,2,...,$ height:
for $j = 1,2,...,$ width:

$$k = R_1$$
$$r_{ij} = I_m(i,j,1)$$
$$g_{ij} = I_m(i,j,2)$$
$$b_{ij} = I_m(i,j,3)$$
$$d_c = \frac{2b_{ij} - 2g_{ij}}{3}$$
$$d_l = \frac{(r_{max} - r_{ij}) + (g_{max} - g_{ij}) + (b_{max} - b_{ij})}{3}$$

if $d_c < 0.05$ and $d_l < 0.05$ then:

$$R_3 = [1,2,3]$$

if $R_2 = 1$ then:

$$I_m(i,j,1) = B_{ijk}(i,j,1)$$
$$I_m(i,j,2) = B_{ijk}(i,j,2)$$
$$I_m(i,j,3) = B_{ijk}(i,j,3)$$

if $R_2 = 2$ then:

$$I_m(i,j,1) = B_{ijk}(i,j,3)$$
$$I_m(i,j,2) = B_{ijk}(i,j,2)$$
$$I_m(i,j,3) = B_{ijk}(i,j,1)$$

if $R_2 = 3$ then:

$$I_m(i,j,1) = B_{ijk}(i,j,2)$$
$$I_m(i,j,2) = B_{ijk}(i,j,1)$$
$$I_m(i,j,3) = B_{ijk}(i,j,3)$$

Output: I_m ■

6.3.2 STRUCTURE OF THE CNN

The CNN's structure definition starts by resizing the dataset images from 300×300 pixels to 128×128 pixels to reduce the problem size, maintaining images in recognisable sizes.

After that, we prepare the first layer of our proposal using the Google MobileNet2 as base structure and the dataset image dimensions with RGB representation.

TABLE 6.1

Layers, Shape and Parameters in the Proposed CNN's Structure

Layer	Output Shape	Number of Parameters
MobileNet2	4,4,128	2,257,984
2D Global Average Pooling	1,1280	0
Dense SoftMax	1,3	3,843
	Total parameters	2,261,827
	Trainable parameters	2,227,715

The MobileNet2 used is included without the top classifying layer because we proposed improving the classification accuracy instead of using transfer learning. Thus, we retrained all the trainable parameters of the MobileNet2 to adapt them with our proposed data augmentation.

After that, we use a 2D global average pooling layer to summarise the information obtained from the Google MobileNet2; this helps reduce underfitting in the following stages.

Finally, we include a dense prediction layer with SoftMax as the activation function, which classifies the three types of signals and return outputs in ranges [0,1] corresponding with 0%–100% of reliability for the hand sign identified.

Table 6.1 shows the shape and trainable parameters in the MobileNet2, the 2D global average pooling layer and the prediction dense SoftMax layer, which built the CNNs structure used in this work.

Figure 6.7 shows the general structure of CNN used in this work.

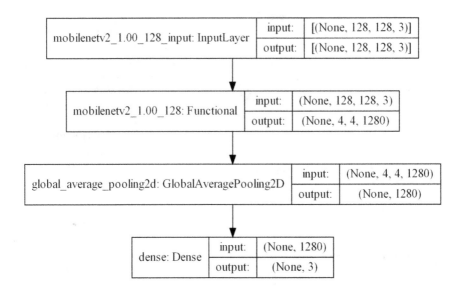

FIGURE 6.7   Layer of proposed CNN with basis on Google MobileNet2.

6.3.3 Data Augmentation for Training and Validation

The CNN's structure in Section 6.3.2 and training implementation uses Python™
with the Google TensorFlow library, including the transformations for data augmentation defined in Section 6.2.3.

In this work, we additionally implemented as augmentation operation the proposed changing color transformation defined in Section 6.3.1.

The data augmentation for training starts by sampling the training images. Then
images are transformed with random values for brightness, rotation, translation, shear,
vertical flip, horizontal flip, zooming and the proposed background color transformation.

Finally, we also transform the validation images and testing images with the same
data augmentation operations to score the proposed hand signs classifier's accuracy,
based on the dataset augmented.

Table 6.2 shows the augmentation operations and their corresponding parameters
to explain their selection.

6.3.4 Training CNN

After augmenting the original dataset for the sampling and training sets, we train
the CNNs in the Tensorflow Google™ library using the categorical cross-entropy in
Equation (6.10) as loss function, which measures the error in categorical distribu-
tions based on SoftMax predicting output functions [31].

TABLE 6.2

**Augmentation Operations Used the Set Configuration Parameters and Their
Explanation**

Augmentation Operation	Configured Parameters	Explanation
Brightness	0.5–1.3	It covers the visible conditions for perceived the hand with low-to-high luminosity
Rotation	0–360°	It allows perceiving the hand signs in every possible orientation
Translation	0–0.06	It allows a variation of 6% of translation; this forces the user to put its hand in the center for giving instructions
Shear	0–0.1	It allows a tilt variation of up to 10% for perceiving the hand signals
Horizontal-vertical flip	True	Allows flipping of hand so that right and left hands are recognised, depending on the users' preference
Zoom-scaling range	0.8–1.5	Accepts hand size variation from 80% to 150% of the dataset size, allowing different users to be recognised
Proposed change background	True, 0.6	Allow changing the background with 60% probability using the proposed algorithm and considering industrial environments

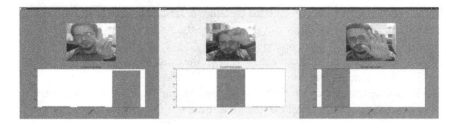

FIGURE 6.8 Designed interface for transmitting recognised orders with audio, text and color.

$$\mathcal{L}_{xe}(\mathbf{y},\mathbf{z}) = -\sum_{i=1}^{K} y_i \log \sigma(z_i) \qquad (6.10)$$

where $\mathbf{y}$ is the target output array, $\mathbf{z}$ is the output array of the predicting layer in the CNN, and $\sigma(z_i)$ is the SoftMax evaluation for the i sample of the dataset.

We set the accuracy as the evaluation metric given by $CA = TCS / TS$, where TCS are the true classified samples and TS are the total samples, as described in Ref. [32].

Then, we trained the CNN using the Adam approach in Algorithm 6.1 described in Section 6.2.2.3, starting with a learning rate $\partial = 1.0E - 4.0$ that decays 10% after two epochs without improving accuracy validation. We allowed early stopping after three epochs without cross-validated accuracy improvement.

6.3.5 CLASSIFICATION SYSTEM FOR TRANSMITTING TEXT AND AUDIO ORDERS

We developed an interface that acquires images from a digital camera using the OpenCV library for Python.

Then, based on the recognised hand signals with the trained CNN, it transmits the orders by reproducing the audio responses "Emergency Stop," "Warning Event," or "Run Process," using the Microsoft Voices by using win32 libraries.

Every order recognised also changes a text label with the recognised instruction and the interface's color to red, yellow or green, corresponding respectively with paper, rock and scissors hand signs, as shown in Figure 6.8.

6.4 RESULTS AND DISCUSSION

6.4.1 DESIGN OF EXPERIMENT

All the experiment stages used a Windows 10 desktop computer with processor Intel(R) Core i7-6700 CPU 3.40 GHz, 16.0 GB RAM, NVIDIA GeForce GTX 970 graphics card, Logitech HD 1080p camera as image acquisition system, TensorFlow 2.4 and Python 3.7.

We start by augmenting transforming all the images in the original dataset obtained from [18], which initially contains 2,892 images. After data augmentation with operations described in Section 6.3.3, we obtained 26,856 images distributed in 25,752 images for training, 732 images for validation and 372 for testing.

Then, we trained the CNN with the structure described in Section 6.2.2 during five epochs, with a batch size of 36 images and using the Adam train in Algorithm 6.1 and the parameters described in Section 6.3.4.

After that, we evaluated our classifier with the accuracy metrics evaluated using the 372 images in the test set.

Finally, we tested the interface with text, audio and color corresponding instructions depending on the digital camera's real-time acquired images.

6.4.2   RESULTS

6.4.2.1   Changing Background Function

The proposed changing background function modified the background characteristics for the training, validating and testing samples in the dataset.

Figure 6.9 shows 25 images sampled from the dataset with background transformation using a 60% probability together with the augmentation operations described in Section 6.3.3.

FIGURE 6.9   Dataset samples with data augmentation, including proposed background change.

TABLE 6.3
CNN Training Results after Five Epochs

Epoch	Loss	Loss Validated	Accuracy	Accuracy Validated
1	0.2718	0.5242	0.8910	0.9078
2	0.0471	0.1206	0.9847	0.9797
3	0.0336	0.0727	0.9892	0.9844
4	0.0267	0.2031	0.9911	0.9688
5	0.0258	0.1101	0.9916	0.9750

6.4.2.2 Training Results

Table 6.3 shows loss value, loss validated, accuracy and accuracy validated per epoch; for the five epochs trained. The third epoch is bold because it is the conserved model since it has the best accuracy validated.

Figure 6.10 shows the loss (blue signal) and the validated loss (orange signal) across the five epochs of training, showing that the lower loss is in the third epoch. Figure 6.11 compares the accuracy (blue signal) and validated accuracy (orange signal) across epochs, showing that the third epoch reaches the best accuracy.

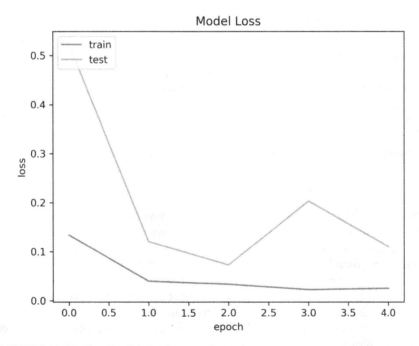

FIGURE 6.10   Loss and validation loss across epochs.

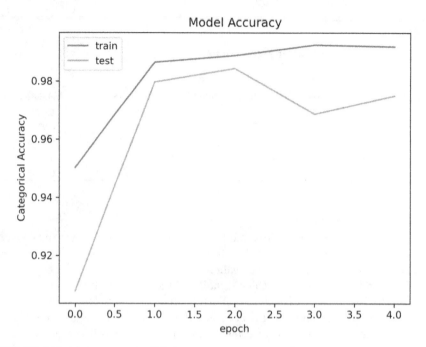

FIGURE 6.11   Accuracy and validation accuracy across epochs.

6.4.2.3 Testing Results

After training the CNNs, we evaluated the 372 testing samples without using the background transformation. The model reaches 0.9543 of accuracy or 95.43% with a loss value of 0.1877. Figure 6.12 shows 25 random samples without background transformation of the test set with their corresponding prediction.

We then evaluated the 372 testing samples but used the background transformation. The model reaches 0.9462 of accuracy or 94.62% with a loss value of 0.2394. Figure 6.13 shows the 25 random samples of the test set with background changing and corresponding prediction.

6.5 CONCLUSION

This chapter presents a novel assistant for transmitting orders with text and audio of deaf-mute supervisors in Industry 4.0. The proposed assistant uses a digital camera that obtains images for identifying hand signs associated with the manufacturing commands "Stop Process," "Warning Event," "Run Process."

Our classifier based in a MobileNet2 CNN uses a dataset of the famous game rock, paper and scissors [18] with 2,892 images, but using our proposal background transformation increasing the dataset to 26,856 images.

The CNN trained reaches 95.43% accuracy in the testing set without background transformation and 94.62% with the proposed background transformation.

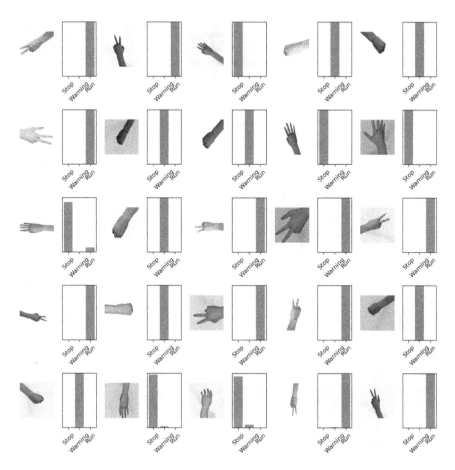

FIGURE 6.12 Predictions of 25 random samples of the testing set without background transformation.

The final implementation processes an image and adjusts the interface in 0.15326 seconds or achieves 6.5248 frames per second, which is enough, especially if it is considered the required time for transmitting the audio order.

6.5.1 Future Work

Although the proposed assistant can transmit three fundamental manufacturing orders that allow interaction between workers and deaf-mute supervisors, these orders are not enough to express all the required interactions in a manufacturing area. Thus, training new signs or a complete system for recognising more instructions will improve the proposed assistant. However, any change must consider the processing speed because the delay in the transmission of orders could result in unsafe conditions.

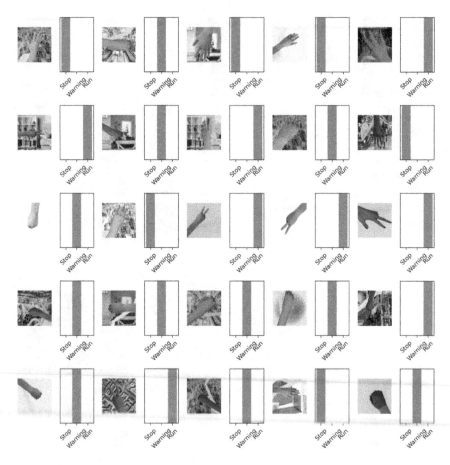

FIGURE 6.13 Predictions of 25 random samples of the testing set with background transformation.

REFERENCES

1. Grodek-Szostak, Z., Siguencia, L. O., Szelag-Sikora, A., & Marzano, G. (2020). The impact of Industry 4.0 on the labor market. 2020 61st International Scientific Conference on Information Technology and Management Science of Riga Technical University, ITMS 2020 – Proceedings. Institute of Electrical and Electronics Engineers Inc. https://doi.org/10.1109/ITMS51158.2020.9259295

2. Emeric, C., Geoffroy, D., & Paul-Eric, D. (2020). Development of a new robotic programming support system for operators. Procedia Manufacturing, 51, 73–80. https://doi.org/10.1016/j.promfg.2020.10.012

3. Villar, L. M., Oliva-Lopez, E., Luis-Pineda, O., Benešová, A., Tupa, J., & Garza-Reyes, J. A. (2020). Fostering economic growth, social inclusion & sustainability in Industry 4.0: A systemic approach. Procedia Manufacturing, 51, 1755–1762. https://doi.org/10.1016/j.promfg.2020.10.244

4. Fatorachian, H., & Kazemi, H. (2021). Impact of Industry 4.0 on supply chain performance. Production Planning and Control, 32(1), 63–81. https://doi.org/10.1080/09537287.2020.1712487

5. Goodley, D., Cameron, D., Liddiard, K., Parry, B., Runswick-Cole, K., Whitburn, B., & Wong, M. E. (2020). Rebooting inclusive education? New technologies and disabled people. Canadian Journal of Disability Studies, 9(5). https://doi.org/10.15353/cjds. v9i5.707

6. da Rosa Tavares, J. E., & Victória Barbosa, J. L. (2020). Apollo SignSound: An intelligent system applied to ubiquitous healthcare of deaf people. Journal of Reliable Intelligent Environments, 1–14. https://doi.org/10.1007/s40860-020-00119-w

7. Stokar, H. (2020). Reasonable accommodation for workers who are deaf: Differences in ADA knowledge between supervisors and advocates. JADARA, 53(2), 38–59. Retrieved from https://repository.wcsu.edu/jadara/vol53/iss2/2

8. Vaidya, O., Gandhe, S., Sharma, A., Bhate, A., Bhosale, V., & Mahale, R. (2020). Design and development of hand gesture based communication device for deaf and mute people. 2020 IEEE Bombay Section Signature Conference (IBSSC) (pp. 102–106). IEEE. https://doi.org/10.1109/IBSSC51096.2020.9332208

9. Lee, C.-C., & Gao, Z. (2020). Sign language recognition using two-stream convolutional neural networks with Wi-Fi signals. Applied Sciences, 10(24), 9005. https://doi.org/10.3390/app10249005

10. Saggio, G., Cavallo, P., Ricci, M., Errico, V., Zea, J., & Benalcázar, M. E. (2020). Sign language recognition using wearable electronics: Implementing K-nearest neighbors with dynamic time warping and convolutional neural network algorithms. Sensors (Switzerland), 20(14), 3879. https://doi.org/10.3390/s20143879

11. Jiang, D., Zheng, Z., Li, G., Sun, Y., Kong, J., Jiang, G., ..., Ju, Z. (2019). Gesture recognition based on binocular vision. Cluster Computing, 22(6), 13261–13271. https://doi.org/10.1007/s10586-018-1844-5

12. Islam, J., Ghosh, A., Iqbal, M. I., Meem, S., & Ahmad, N. (2020). Integration of home assistance with a gesture controlled robotic arm. 2020 IEEE Region 10 Symposium, TENSYMP 2020 (pp. 266–270). Institute of Electrical and Electronics Engineers Inc. https://doi.org/10.1109/TENSYMP50017.2020.9230893

13. Duhan, M., Gulati, U., & Ishaan. (2020). Intelligent system to make the world hear DeafMute people. Proceedings of 3rd International Conference on Emerging Technologies in Computer Engineering: Machine Learning and Internet of Things, ICETCE 2020 (pp. 46–51). Institute of Electrical and Electronics Engineers Inc. https://doi.org/10.1109/ICETCE48199.2020.9091775

14. Chen, L., Fu, J., Wu, Y., Li, H., & Zheng, B. (2020). Hand gesture recognition using compact CNN via surface electromyography signals. Sensors, 20(3), 672. https://doi.org/10.3390/s20030672

15. Li, Q., Lin, Y., & He, W. (2021). SSD7-FFAM: A real-time object detection network friendly to embedded devices from scratch. Applied Sciences, 11(3), 1096. https://doi.org/10.3390/app11031096

16. Tran, D.-S., Ho, N.-H., Yang, H.-J., Baek, E.-T., Kim, S.-H., & Lee, G. (2020). Real-time hand gesture spotting and recognition using RGB-D camera and 3D convolutional neural network. Applied Sciences, 10(2), 722. https://doi.org/10.3390/app10020722

17. Wang, J., Liu, T., & Wang, X. (2020). Human hand gesture recognition with convolutional neural networks for K-12 double-teachers instruction mode classroom. Infrared Physics and Technology, 111, 103464. https://doi.org/10.1016/j.infrared.2020.103464

18. Moroney, L. (n.d.). Rock Paper Scissors Dataset – lmoroney@. Retrieved February 3, 2021, from http://www.laurencemoroney.com/rock-paper-scissors-dataset/

19. Bovik, A. C. (2005). Handbook of image and video processing. 2nd ed. Amsterdam: Elsevier Academic Press.

20. Ibraheem, N. A., Hasan, M. M., Khan, R. Z., & Mishra, P. K. (2012). Understanding Color Models: A Review. Retrieved from http://www.ejournalofscience.org/Download_April_pdf_21.php

21. McGrath, J. R., Beck, M., & Hill, M. E. (2017). Replicating red: Analysis of ceramic slip color with CIELAB color data. Journal of Archaeological Science: Reports, 14, 432–438. https://doi.org/10.1016/j.jasrep.2017.06.020

22. Mokrzycki, W., & Tatol, M. (2011). Color difference delta E – A survey colour difference Δ E – A survey faculty of mathematics and informatics. Machine Graphics and Vision, 20(4), 383–411.

23. Goodfellow, I., Bengio, Y., & Courville, A. (2016). Deep Learning. MIT Press. Retrieved from http://www.deeplearningbook.org

24. Gu, J., Wang, Z., Kuen, J., Ma, L., Shahroudy, A., Shuai, B., …, Chen, T. (2018). Recent advances in convolutional neural networks. Pattern Recognition, 77, 354–377. https://doi.org/10.1016/j.patcog.2017.10.013

25. Bouchard, G. (2007). Efficient Bounds for the Softmax Function and Applications to Approximate Inference in Hybrid models.

26. Kingma, D. P., & Ba, J. L. (2015). Adam: A method for stochastic optimization. 3rd International Conference on Learning Representations, ICLR 2015 - Conference Track Proceedings. International Conference on Learning Representations, ICLR. Retrieved from https://arxiv.org/abs/1412.6980v9

27. Abadi, M., Barham, P., Chen, J., Chen, Z., Davis, A., Dean, J., …, Zheng, X. (2016). TensorFlow: A system for large-scale machine learning. Proceedings of the 12th USENIX Symposium on Operating Systems Design and Implementation, OSDI 2016 (pp. 265–283).

28. Shorten, C., & Khoshgoftaar, T. M. (2019). A survey on image data augmentation for deep learning. Journal of Big Data, 6(1), 60. https://doi.org/10.1186/s40537-019-0197-0

29. Hussain, Z., Gimenez, F., Yi, D., & Rubin, D. (2017). Differential data augmentation techniques for medical imaging classification tasks. AMIA annual symposium proceedings, AMIA symposium, 2017 (pp. 979–984). Retrieved from/pmc/articles/PMC5977656/

30. Taylor, L., & Nitschke, G. (2019). Improving deep learning with generic data augmentation. Proceedings of the 2018 IEEE symposium series on computational intelligence, SSCI 2018 (pp. 1542–1547). Institute of Electrical and Electronics Engineers Inc. https://doi.org/10.1109/SSCI.2018.8628742

31. Li, V., & Maki, A. (2019). Feature contraction: New convnet regularization in image classification. British Machine Vision Conference 2018, BMVC 2018 (pp. 1–11).

32. Vihinen, M. (2012). How to evaluate performance of prediction methods? Measures and their interpretation in variation effect analysis. BMC genomics, 13(4), S2. https://doi.org/10.1186/1471-2164-13-S4-S2

7 A Deep Learning Approach to Classify the Causes of Depression from Reddit Posts

Ankita Biswas, Arion Mitra, Ananya Ghosh,
Namrata Das, Nikita Ghosh and
Ahona Ghosh

CONTENTS

DOI: 10.1201/9781003122401-7

7.1   INTRODUCTION

Machine vision has been an emerging paradigm in the last decade for its applicability in a wide range of domains in Industry 4.0. Human lives are surrounded by
two main parts, one is about their physical well-beings and the other is about their
mental state. The turbulence in the first one can be detected in much easier ways
than in the latter. Among various mental illnesses, depressive disorder or depression is the most common one. It is a psychological disease that affects directly
on the personal feelings and negatively leaves a signature in the atmosphere and
generates a state of anxiety. As an immediate result, people may suffer from sadness and loss of interest in things or activities that they used to like or prefer to do
previously.

7.1.1   REASONS BEHIND DEPRESSION

There exist a lot of reasons behind depressive trauma. Some of the most noticeable
reasons are summarised as follows.

7.1.1.1   Abusive Experience

Previous mental, physical, sexual, emotional abuse even domestic violence can lead
to serious depression in one's life. Any shocking experience can affect the mind
permanently and can trigger anxiety from time to time.

7.1.1.2   Loss or Death

Permanent loss of something special or the death of some near and dear ones can
cause serious depression. Even if the loss or the death happened normally, the
affected person can't get over that.

7.1.1.3   Crucial Events

Some bad, or even some good events too, can contribute to the risk of depression.
Some good consequences like promotion, joining a new job, completing a degree
can produce depressive disorder. On the other hand, losing a job, family crisis are
important origins of dejection.

7.1.1.4   Loneliness

Staying alone may generate a great risk in improving mental disorders. Sometimes, loneliness for a little amount of time, hardly for 1 or 2 hours, can increase the degree of depression too [Klein, D.N., & Santiago, N.J. 2003]. The younger generation seems to be more susceptible to higher extents of depression, especially during the COVID-19 pandemic's enforced quarantine periods or lockdown.

7.1.1.5   Medications

The risk of depression can increase by using some medications, such as antiviral drug interferon-alpha, isotretinoin. Taking such medicines for a long period can be harmful.

7.1.1.6   Addiction

Several types of addictions affect the mental health of human beings, be it alcohol, drugs or some other materials. Nowadays, addiction to social media is a strong point for developing severe depression among youngsters. The hallucination of the real world can lead to huge mental manipulation, and as a result it may cause grief, sadness, anxiety, violence to take a permanent place in the addicted mind.

7.1.1.7   Gun Violence

Gun violence can affect people to some extent that may lead to post-traumatic stress disorder. The effect of PTSD extends to the witnesses as well as the survivors of the mishap, including their families and friends [Sharma, S. 2017]. Studies show this kind of mental health disorder may lead to severe depression also.

For the above-mentioned causes of depression, we have classified the underlying causes into five categories, economic difficulties, isolation, substance use, gun violence and domestic stress. Based on these classifications, we have tried to classify the causes and if the post has any signs of suicidality. These predictions are done by using neural network architectures. Long short-term memory (LSTM) and convolutional neural network (CNN) models were trained using word embeddings from pretrained models of GloVe, FastText and Word2Vec. The dataset was divided into two parts, the training dataset and the testing dataset, respectively. Every row has a corresponding post. One of the columns finds any occurrence of suicide in post or not and the other five columns denote the causes of depression for each post. After training the models, we can predict and analyse the cause of depression. Different rehabilitative techniques have been applied in a different state of the art to restore someone's physical and cognitive ability to its original state [Saha, S., & Ghosh, A. 2019] and sentiment analysis of social media posts has played a vital role in different related works [Biswas, S. et al. 2020].

The next section will discuss some recent works conducted in the concerned domain. Section 7.3 will describe the proposed methodology. Section 7.4 will present the outcome of the model and compare the performances of two models having four variations in each. The concluding statements and a possible road map for the future will be presented in Sections 7.5 and 7.6.

7.2   BACKGROUND STUDY

The very first occurrence of depression can be found in the literature search of Medline in the years 1976–1996. Almost in the same period, that is, in the years 1982–1996, a similar search can be discovered. Similarly, from 1985 to 1996 and from 1990 to 1996, two more research approaches were advised [Antoni, M.H. et al. 2001]. Throughout this period, several different diagnostic parameters were improved. The Diagnostic and Statistical Manual of the American Psychiatric Association, which is known as DSM-IV, was founded in the year 1994 [Antoni, M.H. et al. 2001]. It elaborates two different subparts of depression:

1. MDD or Major Depressive Disorder
    Here, MDD is classified as one or many depressive states and the absence of manic states. Depressive mood, loss of interest in regular work and activities are the symptoms of MDD. If a number of these symptoms are present in one's life for nearly about 14–15 days, the person is stated as a depressive patient [Antoni, M.H. et al. 2001]. MDD is observed more in women (12.2%) compared to men (5.3%). Sorrow, indecision as well as self-criticism, irritability were observed predominantly in women [Loades, M.E. et al. 2020].
2. DD or Dysthymia or Dysthymic Disorder
    If the symptoms are not all of enough severity, it is dysthymia [Smith, M.F. et al. 2020]. Here, the criteria are not satisfied called a MDD. But, here also, the patients suffering from dysthymia need to be taken utmost care [Antoni, M.H. et al. 2001]. In 1991, almost 15,000 adults were surveyed in 6 months, and 3% of the population were suffering from depression, and this was recorded as the prevalence of depression [Antoni, M.H. et al. 2001]. In the year 1992, 17 studies were revised and it could be shown that, in one year, the prevalence rate of MDD was between 2.6% and 6.2%. For dysthymia, this rate was 2.3%–3.7% [Antoni, M.H. et al. 2001]. The prevalence rate for lifetime varies from 4.4% to 19.5%. In the year 1994, the National Comorbidity Survey estimated the prevalence rate of MDD as 4.9%. It was 17.9% for lifetime prevalence. This survey took almost 8098 adults between the ages of 18 and 54, and it continued for 30 days [Antoni, M.H. et al. 2001]. Using the random effect model, research of some well-known databases such as Medline (the year 1966–2005), Psych Info (the year 1967–2005) and Cinahl (the year 1982–2005) tells the following facts [Sellick, S.M., & Crooks, D.L. 1999]:
3. Specificity: 83.7% (95% CI: 77.5–90.0)
4. Sensitivity: 36.4% (95% CI: 27.9–44.8)
5. Odd ratios: 4.0 (95% CI: 3.2–4.9)

The meta-regression analysis of this survey shows that the age of the samples, the method of the study and the date of the evaluation have a high role on summary sensitivity and the odds of recognition [Sellick, S.M., & Crooks, D.L.

1999]. The rate of the treatment of psychological disorders has hugely improved in the last few years, but still, many cases are unrevealed till now. The actual reasons behind some crucial cases remain undisclosed [Cepoiu, M. et al. 2008]. There are a lot of studies regarding mental disbalance, which are caused by using social media more than it is required. Using social media such as Facebook, Twitter, Reddit and Instagram, the depressive users can be identified by screening surveys, such as their public posts, their opinions on some facts and pictures as well as details [Cepoiu, M. et al. 2008]. According to studies, it can be stated that depressed Twitter users are intended to post something containing negative thoughts, negative vibes toward the world. It is interrelated and works like poison for society. It is shown that almost 15% of social media users are suffering from depression currently [Koo, J. et al. 2017]. A random race between oneself and the others on social media increases the grief, the anxiety in one's mind. It can be even severe for some cases that the affected person becomes vulnerable to step toward suicidal thoughts. Until now, almost 2 million tweets from 476 users were detected as depressed patients clinically, as collected by De Chowdhury [Koo, J. et al. 2017].

7.2.1 USAGE OF CLASSIFIERS

The behavioral attribute is used to cultivate and know the history of risk of depression. Attributes are related to social bindings, language and linguistic styles, emotions, ego projections, conversations about antidepressant medications with one another, etc. [Koo, J. et al. 2017]. Depending upon these attributes, a classifier is built that is designed to give the estimation of risk factor. This classifier is called a support vector machine (SVM) classifier. The accuracy of this classifier is almost 70%. It was discovered by Tsugawa that the word frequency, sentence construction, topic to talk are useful parameters to predict the risk factor of depression [Koo, J. et al. 2017]. Using a question-answer round among 209 participants, the Radial Kernel SVM classifier has become 69% accurate in predicting the existence of depression in 81 participants [Koo, J. et al. 2017].

As well, Reece built the predictive model depending upon some different parameters like the context of the tweets, usage of punctuation marks in appropriate positions, the width of the tweets, usage of slang, etc. Using this model, depressive patients could be successfully distinguished from healthy persons [Koo, J. et al. 2017]. Among 204 depressed candidates, information of almost 105 patients was collected and evaluated. Using a 1200-tree random forest classifier, the best classifier performance was recorded. The accuracy was almost 0.8666, which is much more than other studies [Koo, J. et al. 2017]. Another approach was the "Bag of Words" approach, in which the frequency of a word presented in a tweet is counted, and depending upon this frequency, the level of depression is detected [Swinger, N. et al. 2019]. Four types of binary classifiers are incorporated [Koo, J. et al. 2017]:

1. Decision tree (DT)
2. Naive Bayes (NB) algorithm

3. Linear SVM classifier

4. Logistic regression approach

Among these different classifiers, the Naive Bayes (NB) algorithm performs with the utmost perfection. The accuracy of the NB classifier is almost 81% and the precision is 0.86 [Koo, J. et al. 2017]. The logic of this classifier is based on the Bayes theorem of probabilistic statistics. If and only if the features are independent of each other, the NB theorem can be applied [Swinger, N. et al. 2019]. The Bayes theorem states that,

$$P(X \mid Y) = \frac{P(X \mid Y)P(X)}{P(X)} \tag{7.1}$$

By using the Bayes theorem efficiently, the essence of some different features can be merged. Because of this advantage, the Bayes theorem is highly used in machine learning [Koo, J. et al. 2017]. The frequency of observance of some evidence generates an outcome, and the surety that any particular evidence is happening creates the conclusion. So, it can be stated that the NB classifier is the most desirable classifier technique [Koo, J. et al. 2017].

The decision tree (DT) classifies the samples using a sorting technique on the feature values. Each feature is represented by a single node in the DT and each value that the node can take is represented by each division [Vinchurkar, D.P., & Reshamwala, A. 2012]. It requires the partitioning of data into multiple subdivisions. These subdivisions may contain some similar outcomes [Koo, J. et al. 2017]. Split selection is an important technique used in the DT, which can find the attribute for each test node and the related splitting functions.

In a high-dimensional space, SVM can be applied to underline two different classes. It is supervised as a learning model that can regulate several features keeping the performance balanced. It is a highly appreciable learning model, as it can reduce the possibility of overfitting efficiently [Mowery, D., Bryan, C., & Conway, M. 2017]. SVM works upon the algorithm which takes training data as a labeled example for any binary problem. It is more demanding when working with real-world problems consisting of real datasets. The underlying algorithm of SVM produces a potential hyperplane [Ramírez-Cifuentes, D., & Freire, A. 2018] that distributes the cases of two different classes. The interval between the divided hyperplane and the dataset used for training is increased rapidly by this algorithm [Podgorelec, V., & Zorman, M. 2015]. For a selected object, SVM can predict the side on which that object stays of the hyperlink.

7.2.2 MACHINE VISION AND DEPRESSION DETECTION

Machine vision is the technology used by computers to visualise the environment using image data to inspect and analyse applications such as automatic inspection, process control and robot guidance, usually in industry.

When machine vision methods are used for detection of such mental health problems, we will be able to detect the person's emotion from his facial expressions as well as his body language as a whole i.e., if the person is potentially depressed or not, however, we can't determine the possible causes behind it without any context that can be obtained by counseling him/her or by knowing his account/history, which is essential for recovery. For analysing the cause behind depression, machine vision method shave to be incorporated along with audio extraction from video datasets. The extracted audio can then be converted into text on which we can execute our proposed model to determine the cause.

We have limited our scope to work with text data alone because integrating machine vision methods would require higher computing power for training the data as we need to analyse videos. Moreover, video datasets that would suffice to our goal of detecting depression and determining the possible cause behind it are difficult to obtain on the Internet. Potential datasets can be therapy sessions of clinically depressed individuals, which is tough to get as such datasets are meant to be confidential.

7.2.3 COMPARATIVE STUDY AMONG EXISTING WORKS

Table 7.1 summarises existing works that have been done on depression detection using machine vision methods and Table 7.2. presents a comparative analysis among existing works.

7.3 PROPOSED METHODOLOGY

In this section, we will discuss the different models of deep neural networks and embedding techniques.

7.3.1 NEURAL NETWORK

A neural network model was inspired [Guntuku et al. 2017] from the sophisticated structure of human brains, inside which information is processed parallel through hundreds of billions of neurons that are interconnected to each other. The neural network model used in machine learning, as shown in the Figure 7.2, typically contains one input layer, some hidden layers followed by the output layer of neurons [Swinger, N. et al. 2019]. Based on the depth of the hidden layers, the model is called a deep neural network (more than three layers) or shallow neural network (one or two layers). The generalised equation of a neural network model can be written as,

$$h = \sigma(W1x + b1)$$
$$y = \sigma(W2h + b2)|$$
(7.2)

where h is the hidden layer, x is the input layer, y is the output layer, $b1$, $b2$ are the biases, $W1$, $W2$ are the weights.

7.3.1.1 Recurrent Neural Network (RNN)

Recurrent neural network (RNN) is a specialised algorithm designed for a sequence of data such as $x(t) = x(0), ..., x(t)$ with the time step index t ranging from 1 to n. RNN is a well built and sturdy neural network. The presence of internal memory makes

TABLE 7.1

Comparative Study among Existing Works Using Machine Vision

Refs.	Objective	Methods Used	Result	Benefits	Loopholes
Parameswaran, N.S., and Venkataraman, D. (2019)	A computer vision–based image processing system for depression detection among students for counseling	Gabor filter for facial feature extraction SVM classifier	Facial features are classified from the entire video and then the level of depression is determined by the number of negative emotions present.	Accuracy: 64.38%	First 160 frames of the video are only considered. Recent videos are only considered. Other parameters such as grades, attendance, etc., are not considered.
Venkataraman, D., and Parameswaran, N.S. (2018)	Extraction of facial features for depression detection among students	Viola Jones face detection algorithm Gabor filter bank of 40 filters SVM classifier (multilevel)	Level of each feature (happy, contempt, disgust) is determined for each frame in a video. The measure of presence/absence of happy features describes the amount of negativity and hence the level of depression.	Depending on the level of negativity the required counseling can be recommended	First 50 frames are only considered. Recent videos are only considered. Other parameters such as grades, attendance, etc., are not considered
Dadiz, B.G., and Ruiz, C.R. (2019)	Detecting depression in videos using uniformed local binary pattern on facial features	Keyframe extraction method LBP (local binary pattern) transformation SVM with RBF kernel	Uniformed LBP features are extracted from keyframes of the videos and then classified using SVM to find if a person is depressed or not.	98% Accuracy with SVM with RBF kernel	Training time for neural networks is more. Noise from the video background. PCA (principal component analysis) caused a loss in the model's accuracy.
Cohn, J.F., Kruez, T.S., Matthews, I., Yang, Y., Nguyen, M.H., Padilla, M.T., Zhou, F. and De la Torre, F. (2009)	Detecting depression from facial actions and vocal prosody	Manual FACS coding Active appearance modeling (AAM) Pitch Extraction SVM for FACS Logistic regression for voice	The relation between facial and vocal behavior inspected and used detection of depression.	88% accuracy for manual FACS 79% accuracy for AAM	Limited dataset

TABLE 7.2
Comparative Study among Existing Works

Refs.	Objective	Methods Used	Result	Benefits	Loopholes
Antoni, M.H., et al. (2001)	To decrease the prevalence of depression in early-stage breast cancer among women	The neural network is used. Linguistic metadata is used.	CNN is used for embedding as input vectorisation. The final result is obtained by calculating the mean of the metadata.	This mechanism was almost 70% accurate. It can detect the pre-stage of depression.	It is not so able to detect the intensity correctly. It can predict if and only if it is in an early stage.
Mowery, et al., M. (2017)	Interpretable depression detection from social media	Feature networks: Depressive symptoms, ruminative thinking, writing style, sentiments, etc.	Some words and their corresponding frequencies are: Franxiety, 2233 neg, meds, 1229 neu, Medication, 934 pos, disorder, 698 neg, psychiatrist, 382 neu, Adderall, 320 pos, Suicidal, 316 neg, Disability, 145.	Based on the RSDD training set, it used GloVe for embedding word vectors and GRU. Adam optimiser and weighted loss for the imbalance between depressed and control groups.	It can't explain mental disorders (like dementia, schizophrenia and bipolar disorder). This model uses smaller training data as input as it has limited computing power, showing lower performance than the state-of-the-art model.
Islam, M.R., Kabir, M.A., Ahmed, A., Kamal, A.R.M., Wang, H. and Ulhaq, A. (2018)	Depression detection from social media platform like Facebook	Machine learning, decision tree formation is used to detect depression from social media data.	Using various psycholinguistic methods, the efficiency has been evaluated. The rate of accuracy was almost 69%.	It can judge the contents of the public posts on the social media platform. Machine learning approaches increase efficiency a lot.	It can judge only to a limit. Not all the sample data is connected to the real world, which leaves the method somewhat inaccurate.
Guntuku, S.C., Yaden, D.B., Kern, M.L., Ungar, L.H. and Eichstaedt, J.C. (2017)	Detection of depression by social media harvesting: A multimode dictionary learning solution	Data collection, data pre-processing, feature extraction, multimodal depressive dictionary learning: Uni-modal dictionary learning, multimodal dictionary learning joint sparse representation	Depressed users are more likely to post tweets (+44% on average) between 23:00 and 6:00. They have 0.37 positive words and 0.52 negative words in per tweet.	WDL achieved better performance than NB by 10%, MSNL and the presented MDL outperformed WDL by 5%–8%. The MDL method surpassed the WDL method by 3%	It measures only the benchmark depression and well-defined discriminative depression-oriented feature groups.

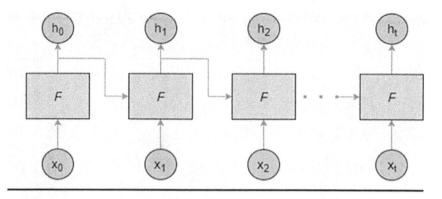

FIGURE 7.1 Illustration of recurrent neural network with the inputs and outputs.

RNN one of the most favorable algorithms in machine learning [Liu, H. et al. 2019]. Because of the presence of internal memory, this model became the first one, to remember its input. This feature of RNN makes it unique and suitable for sequential data. RNN algorithms are regarded as "recurrent" as the same methods are performed for every component of the sequence and the outputs are dependent on the previous calculations.

In Figure 7.1, $x_0, x_1, \ldots, x_t$ are the word embedding vectors, $h_0, h_1, \ldots, h_t$ are the output predictions of the labels, and each F block represents the corresponding neural layer of our RNN model.

7.3.1.2 Long Short-Term Memory (LSTM)

Long short-term memory algorithms are the most powerful, well-known and well-built class of artificial RNN architecture [Zhang, K. et al. 2017]. This is because LSTM is designed to acknowledge patterns in sequences of data. In sequence prediction problems, order dependence is efficiently depicted by LSTM algorithms. This is a behavior required in complex problem domains like machine translation, speech recognition and more. Long short-term memory algorithms are regarded as the complicated field of deep learning.

7.3.1.3 Convolutional Neural Network or ConvNet (CNN)

Processing of images is done using a special multilayered neural network algorithm known as CNN. It is used to detect and analyse objects present in an image [AlSagri, H.S., & Ykhlef, M. 2020]. CNN is a deep learning model designed specially to work with two-dimensional image data. ConvNet requires much less preprocessing as compared to other classification algorithms. Figure 7.2 shows a CNN model with six convolution layers and three output layers.

The building blocks of ConvNet are called filters (or, kernels) [Cho, K. et al. 2014], which are used to take out the significant characteristics from the input using the convolution operation.

Figure 7.3 illustrates the workflow diagram of our deep learning model.

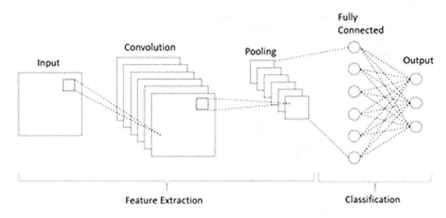

FIGURE 7.2 Structure of CNN.

7.4 EXPERIMENTAL RESULT AND DISCUSSION

In this section, we test the predicted outputs of our deep neural network models and compare their accuracies with each other.

7.4.1 DATASET DESCRIPTION

We have used the Reddit Mental Health dataset [Low, D.M. et al. 2020] which contains posts and text features from 28 mental health and nonmental health subreddits. For our purpose, we have used the CSV files on depression for the years 2018 and 2019. We have merged these two CSV files and will be referred from hereon as a dataset. The dataset contains Reddit posts and corresponding labels denoting the occurrences of the various attributes that are responsible for the depressing post by the user. Figure 7.4 visualises the dataset.

The labels economic_stress_total, isolation_total, substance_use_total, guns_total, domestic_stress_total and suicidality_total contain the number of occurrences of the labels in the post. Our goal is to identify the labels of a post; therefore, we divide the post and labels into *X_train*, *y_train*, respectively.

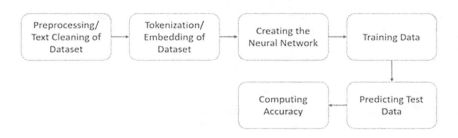

FIGURE 7.3 Complete workflow model.

id		post	economic stress_total	isolation total	substance use_total	guns_total	domestic stress_total	suicidality total
7	7	This is what will happen when I kill myself Th...	2	3	1	0	0	7
8	8	Correlation btw money and mental health We ne...	0	0	1	0	0	0
9	9	The worst part about feding depression is figu...	0	1	1	0	0	3
10	10	Rebuilding walls Hi all, been having on/off th...	0	0	0	0	0	0
11	11	What do I do? I recently changed out of a job...	3	0	0	0	0	0
12	12	Not even the best things make me happy I'm 14...	0	0	0	0	0	0
13	13	I think I have depression? Ive always been den...	0	4	2	0	0	0

FIGURE 7.4 The dataset illustration.

7.4.2 DATA PREPROCESSING

Preprocessing is a necessary step for any textual data that cleans unnecessary items from the text, resulting in better training of the model and thus better prediction of results. There are a lot of preprocessing methods that are followed, the most common among them are tokenisation, stemming, lemmatisation, stop word removal, lower casing, etc. We have used cleaning of text by removing punctuations, numbers, single characters and multiple spaces and stop word removal while using the Word2Vec model. Figure 7.5 illustrates preprocessing of a noisy sample of data.

For the *y_train* values corresponding to the various labels, we observe, instead of binary values, integers denoting the occurrences of that label within the post. For our convenience instead of considering the number of occurrences, we will convert each occurrence to just 1 and no occurrence as 0.

7.4.3 WORD EMBEDDINGS

Word embeddings are dense representations of each word in a text data, considering the context and other surrounding words that the individual word occurs with. One important feature of word embedding is that similar words in a semantic sense have a smaller distance between them than words having no semantic relation. For example, the words *happy*, *sad*, *laughing* should be closer mathematically than the

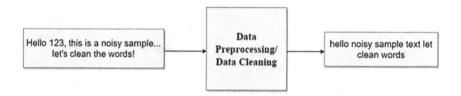

FIGURE 7.5 Cleaning text and stop words removal.

```
[('sad', 0.6913076639175415),
 ('unhappy', 0.5893709659576416),
 ('paranoid', 0.5601510405540466),
 ('anxious', 0.5578929781913757),
 ('lonely', 0.5407476425170898),
 ('faking', 0.5269251465797424),
 ('quite', 0.5263458490371704),
 ('miserable', 0.5195631980895996),
 ('saddepressed', 0.5147014856338501),
 ('frustrated', 0.5111351013183594)]
```

FIGURE 7.6 Semantically similar words to depressed.

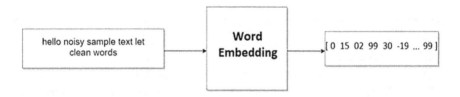

FIGURE 7.7 Embedding process.

words *happy* and *movie*. The output in Figure 7.6 illustrates that using Word2vec [Church, K.W. 2017] embedding model, the semantically closer words to the word *depressed* are listed.

Figure 7.7 illustrates the embedding process of a sample text.

We have used four types of pretrained word embeddings in our program to find out which one performs better.

7.4.3.1 Preprocessing Using Keras

For the first embedding method, we have used Keras preprocessing (text preprocessing) libraries (Keras [Online]). This class vectorises a corpus, by converting each text into either a sequence of integers or into a vector where the coefficient of each token could be a binary value, based on the number of words. Here we have maintained a maximum length of 200 words in the sequence. Figure 7.8 illustrates the embedding vector generated by the Keras library.

```
The embedded vectors of each posts:
[[   0    0    0 ... 1085  142   55]
 [   0    0    0 ...  217   24   25]
 [   0    0    0 ...   61 3506    7]
 ...
 [2226   28    5 ...    6  361 3303]
 [   3 1283   20 ...   21  138 9519]
 [   0    0    0 ...   23  159    7]]
```

FIGURE 7.8 Embedding by Keras library.

```
array([[ 0.          , 0.          , 0.          , ..., 0.          ,
         0.          , 0.          ],
       [-0.18970001,  0.050024   , 0.19084001, ..., -0.39804     ,
         0.47646999, -0.15983    ],
       [-0.071953   ,  0.23127    , 0.023731   , ..., -0.71894997,
         0.86894    ,  0.19539    ],
       ...,
       [ 0.          , 0.          , 0.          , ..., 0.          ,
         0.          , 0.          ],
       [ 0.          , 0.          , 0.          , ..., 0.          ,
         0.          , 0.          ],
       [ 0.          , 0.          , 0.          , ..., 0.          ,
         0.          , 0.          ]])
```

FIGURE 7.9   Embedding by the GloVe model.

7.4.3.2   GloVe Embeddings

GloVe stands for "Global Vectors" [Pennington, J., Socher, R., & Manning, C.D. 2014]. GloVe captures both global and local statistics of a corpus, which means not only does GloVe take into consideration the semantics of surrounding words but also the global context. Figure 7.9 illustrates the embedding vector.

7.4.3.3   Word2Vec Embedding

Word2Vec is similar to an autoencoder that encodes each word in the corpus into a vector and then trains each word against other neighboring words [Zhang, K. et al. 2017]. It comprises two flavors, i.e., Skip-gram and CBOW (common bag of words). Figure 7.10 shows the Word2Vec embedding vector.

7.4.3.4   FastText Embedding

FastText [Bojanowski, P. et al. 2017] is an extension of the Word2Vec model. In this model instead of feeding individual words into the neural network, FastText breaks

```
⌐►  (74344, 128)
    array([[ 0.0000000e+00,  0.0000000e+00,  0.0000000e+00, ...,
             0.0000000e+00,  0.0000000e+00,  0.0000000e+00],
           [ 5.5911730e-01,  1.2055533e+00,  5.7974600e-01, ...,
             2.3710895e+00, -1.9147620e+00,  4.1103268e-01],
           [ 2.4246790e-01, -1.3065898e+00, -1.6430501e+00, ...,
             1.8498662e-01,  1.3990414e-01,  1.0179166e+00],
           ...,
           [ 3.1274013e-02, -3.4905456e-02, -2.3243528e-02, ...,
            -2.0834930e-02,  3.3612435e-03, -1.4485268e-02],
           [ 1.7424286e-03,  7.0311260e-04, -1.0151207e-02, ...,
             1.0940481e-02,  3.0194744e-03, -2.0008232e-02],
           [ 6.2676580e-03, -1.2442744e-02, -2.3057670e-02, ...,
             2.3205564e-03,  6.0968923e-05,  3.0670900e-02]])
```

FIGURE 7.10   Embedding by Word2Vec.

```
preparing embedding matrix...
number of null word embeddings: 23841
array([[ 0.        ,  0.        ,  0.        ,  ...,  0.        ,
         0.        ,  0.        ],
       [-0.27471   , -0.1206    , -0.27090001, ...,  0.01999   ,
         0.074582  ,  0.2022    ],
       [ 0.030817  ,  0.15637   , -0.098392  , ...,  0.087291  ,
         0.18826   , -0.11722   ],
       ...,
       [-0.23138   , -0.77696002, -0.14453   , ...,  0.016665  ,
         0.85293001,  0.11547   ],
       [ 0.        ,  0.        ,  0.        ,  ...,  0.        ,
         0.        ,  0.        ],
       [ 0.        ,  0.        ,  0.        ,  ...,  0.        ,
         0.        ,  0.        ]])
```

FIGURE 7.11   Embedding by FastText.

words into several *n*-grams (sub-words). After training the neural network, word embeddings for all the n-grams from the training dataset will be generated. The representation of rare words becomes efficient since some of its n-grams may be common with some other words. Figure 7.11 shows the FastText embedding vector.

After the embedding matrix is obtained, it is then fed into the neural network for training based on the defined parameters.

7.4.4   IMPLEMENTATION OF DEEP NEURAL NETWORKS

In this part, we define the two types of neural networks CNNs and LSTM. These are described in the following sections.

7.4.4.1   CNN

We have implemented our CNN model using four different kinds of embedding models: Keras preprocessing method, GloVe, Word2Vec, FastText. The model consists of nine layers as shown in Figure 7.12.

1. *Input layer*: All data from the "post" column in the training dataset is fed into this layer after being converted into their corresponding tokenised vectors (using Tokeniser() from Keras). Each of these vectors is of length 200.
2. *Embedding layer*: This layer tries to find the optimal mapping of each vector to a vector of real numbers. Here, each component of the vector is used as an index to access the embedding weight matrix that contains vectors for each word. The vocabulary size of the embedding matrix is specified. The output of this layer is an array of 2D matrices, where the matrices correspond to the sentences in the "post" column. The dimension of these matrices is 200 × 128.
3. *SpatialDropout1D*: This layer is used to encourage independence between feature maps as the adjacent frames within feature maps can be correlated.

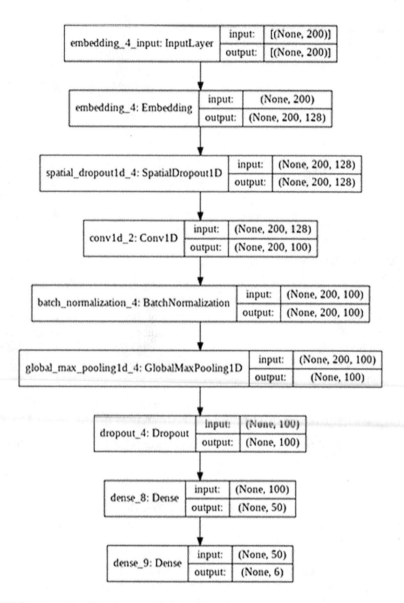

FIGURE 7.12   . The CNN layers with their dimensions.

It drops the correlated sections to avoid learning from unnecessary data. It drops such entire 1D feature maps.

4. *Conv1D*: This layer produces 100 convolutional kernels/filters where each convolution window is of length 4, which is entwined with the layer input over some activation function, ReLu in this case. The dimension of the output changes according to the filter dimension, i.e., 100 in this case.

5. *Batch normalisation*: This layer normalises the inputs into standard form, i.e., they will have a mean of zero and a standard deviation of one.

6. *GlobalMaxPooling1D*: This layer downsises the input by one dimension by taking the maximum of each input.

7. *Dropout*: This layer randomly drops out nodes with a given probability, 0.5 in this case. This helps in avoiding overfitting of data.

8. *Dense of dimensionality 50*: This layer is fully connected. Thus, this layer provides learning features from all the combinations of the features of the previous layer. Finally, the output is passed through the activation function ReLu.

9. *Dense of dimensionality 6*: This layer is the same as the previous layer with lower dimensionality to provide the final output. Our output consists of 6 parameters; thus the dimensionality of this layer has been set to 6.

7.4.4.2 RNN

We have implemented our RNN model using four different kinds of embedding methods: Keras preprocessing method, GloVe, Word2Vec, FastText. The model consists of nine layers which is depicted in Figure 7.13.

This RNN model uses python's library called "FastText" for generating word embeddings. The functionality of the layers in this model is the same as the CNN model except for the fourth layer that consists of a bidirectional LSTM. Bidirectional LSTM trains two LSTMs on the input sequence. First on the input as it is and the second time it reverses the input stream and applies LSTM on it again. This gives clarity in terms of the context.

7.4.5 TRAINING

Before training our model on a given dataset, we need to specify some hyperparameters for it, which will be used to control the learning process. One such hyperparameter is "epoch." Epoch defines the number of times a given machine learning algorithm will act on a given training dataset. It is typically expected from a machine learning algorithm to have multiple epochs. This way we can observe if multiple training sessions result in improvement of the model's accuracy or are deteriorating it. It can also happen that with incrementing epoch a model initially improves and then deteriorates, or vice versa.

After a certain number of epochs of the model, we observed that our model started overfitting and to avoid that we considered another hyperparameter called "patience." We arbitrarily set the epoch to a high value and then set the patience. If it is observed that with increasing epoch the model's accuracy is not showing any visible improvements, then depending on the patience value, the training will stop, in other words, epoch won't be incremented. The higher the patience value, the longer it takes for the training process to stop. Another hyperparameter called "batch size" determines the number of samples to work upon in one iteration. After setting these hyperparameters, we will train the model on the training dataset by passing the input data and the associated labels as parameters.

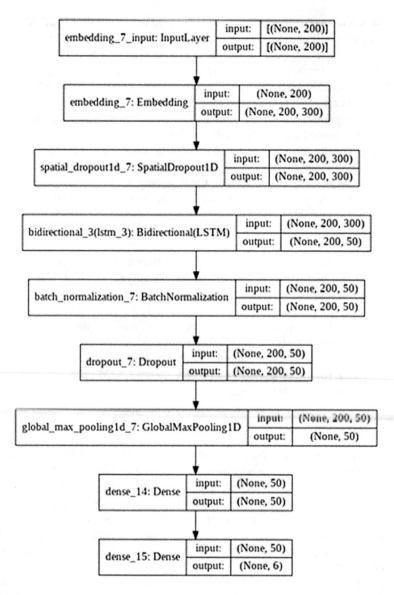

FIGURE 7.13   The LSTM layers with their dimensions.

7.4.6   TESTING

We have a separate dataset called the test dataset on which we executed our model. Then we compared the predicted outputs with the original outputs and calculated the accuracy of our model. We have chosen some random indices from the testing dataset and have executed all of our models on the data corresponding to those indices as shown next.

Post ID	Economic Stress	Economic Stress Prediction	Isolation	Isolation Prediction	Substance Use	Substance Use Prediction
262	1	0.040451463	0	6.75076e-05	0	0.005170262
377	0	0.025747016	0	0.06921837	0	0.015796121
419	0	0.006119883	1	0.9946484	0	0.0011981586
7157	0	0.047943693	0	0.021968199	1	0.99980956
9030	1	0.9993117	1	0.06635439	1	0.06493198

Guns	Guns Prediction	Domestic Stress	Domestic Stress Prediction	Suicidality	Suicidality Prediction
1	0.2834513	0	0.70632243	1	0.9930086
0	0.20134191	1	0.7973459	1	0.99899286
0	0.00013812166	0	0.0044075744	0	0.4235073
0	0.009211633	0	0.06523587	0	0.033608943
0	0.01432148	1	0.2046989	0	0.03792415

FIGURE 7.14   Actual value vs predicted value with CNN using Keras preprocessing.

7.4.6.1   CNN

After training our CNN model with Reddit posts, in this section, our predicted reasons and the actual reasons for depression have been tallied and then we have plotted graphs to visualise the accuracy and the loss of our model.

7.4.6.1.1   CNN Using Keras Preprocessing

Figure 7.14 shows the result of implementing CNN using Keras preprocessing.

Figure 7.15 shows graphs depicting training vs testing accuracy and loss over epoch, respectively, for CNN using Keras preprocessing model.

7.4.6.1.2   CNN Using GloVe

Figure 7.16 shows the result of implementing CNN using GloVe.

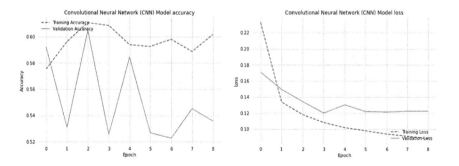

FIGURE 7.15   Graphs depicting training vs testing accuracy and loss over epoch for CNN using Keras preprocessing model.

Post ID	Economic Stress	Economic Stress Prediction	Isolation	Isolation Prediction	Substance Use	Substance Use Prediction	Guns	Guns Prediction	Domestic Stress	Domestic Stress Prediction	Suicidality	Suicidality Prediction
262	1	0.06177452	0	0.030286485	0	0.03782178	1	0.021535365	0	0.023938928	1	0.9890471
377	0	0.06145048	0	0.037565183	0	0.033757124	0	0.037868224	1	0.96585464	1	0.99618345
419	0	0.05649312	1	0.7176083	0	0.015147066	0	0.0025077274	0	0.008057287	0	0.051346704
7157	0	0.48329926	0	0.24254502	1	0.99759066	0	0.028806318	0	0.13741317	0	0.20339793
9030	1	0.5596884	1	0.23319764	1	0.3016379	0	0.029454017	1	0.12610042	0	0.20007996

FIGURE 7.16   Actual value vs predicted value with CNN using GloVe.

Figure 7.17 shows graphs depicting training vs testing accuracy and loss over epoch, respectively, for CNN-GloVe model.

7.4.6.1.3   CNN Using Word2Vec
Figure 7.18 shows the result of implementing CNN using Word2Vec.

Figure 7.19 shows graphs depicting training vs testing accuracy and loss over epoch, respectively, for CNN-Word2Vec model.

7.4.6.1.4   CNN Using FastText
Figure 7.20 shows the result of implementing CNN using FastText.

Figure 7.21 shows graphs depicting training vs testing accuracy and loss over epoch, respectively, for CNN-FastText model.

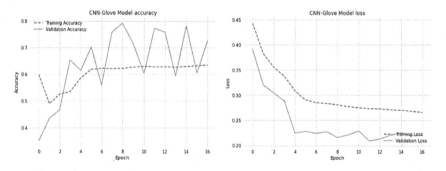

FIGURE 7.17   Graphs depicting training vs testing accuracy and loss over epoch for CNN-GloVe model.

Post ID	Economic Stress	Economic Stress Prediction	Isolation	Isolation Prediction	Substance Use	Substance Use Prediction
262	1	0.011762363	0	0.0021227298	0	0.0089044925
377	0	0.12542532	0	0.028644409	0	0.10796845
419	0	0.056247227	1	0.050775804	0	0.026506534
7157	0	0.1066001	0	0.030244123	1	1.0
9030	1	0.122265406	1	0.028705874	1	0.052041776

Guns	Guns Prediction	Domestic Stress	Domestic Stress Prediction	Suicidality	Suicidality Prediction
1	0.004873513	0	0.0024606006	1	0.99999964
0	0.018571408	1	0.97951186	1	0.9999988
0	0.0040844837	0	0.014184389	0	0.0637706
0	0.010341184	0	0.020721985	0	0.06726594
0	0.008141157	1	0.02943556	0	0.037224233

FIGURE 7.18 Actual value vs predicted value with CNN using Word2Vec.

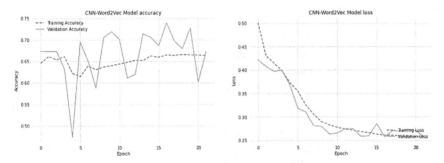

FIGURE 7.19 Graphs depicting training vs testing accuracy and loss over epoch for CNN-Word2Vec model.

Post ID	Economic Stress	Economic Stress Prediction	Isolation	Isolation Prediction	Substance Use	Substance Use Prediction
262	1	0.025825309	0	0.0068893475	0	0.008189455
377	0	0.057231344	0	0.014230417	0	0.025207281
419	0	0.029527077	1	0.999739	0	0.0069020335
7157	0	0.353294	0	0.13937959	1	1.0
9030	1	0.41789103	1	0.15859133	1	0.2610282

Guns	Guns Prediction	Domestic Stress	Domestic Stress Prediction	Suicidality	Suicidality Prediction
1	0.0023227646	0	0.0110708345	1	1.0
0	0.008303354	1	0.34754324	1	1.0
0	0.0016306544	0	0.007718866	0	0.027377155
0	0.012858463	0	0.074575834	0	0.16004637
0	0.013416875	1	0.113433756	0	0.12800156

FIGURE 7.20 Actual value vs predicted value with CNN using FastText.

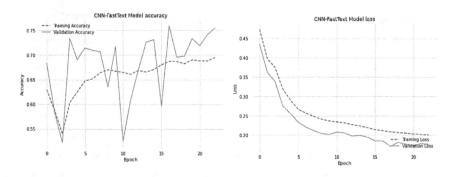

FIGURE 7.21   Graphs depicting training vs testing accuracy and loss over epoch for CNN-FastText model.

7.4.6.2   RNN

After training our RNN model with Reddit posts, in this section, our predicted reasons and the actual reasons for depression have been tallied and then we plotted graphs to visualise the accuracy and the loss of our model.

7.4.6.2.1   RNN Using Keras Preprocessing

Figure 7.22 shows the result of implementing RNN using Keras preprocessing.

Figure 7.23 shows graphs depicting training vs testing accuracy and loss over epoch, respectively, for the RNN using Keras preprocessing model.

7.4.6.2.2   RNN Using GloVe

Figure 7.24 shows the result of implementing RNN using GloVe.

Figure 7.25 shows graphs depicting training vs testing accuracy and loss over epoch, respectively, for RNN-GloVe model.

Post ID	Economic Stress	Economic Stress Prediction	Isolation	Isolation Prediction	Substance Use	Substance Use Prediction
262	1	0.0007867766	0	0.019779798	0	0.0119473
377	0	0.0221459	0	0.05457283	0	0.06056982
419	0	0.019711936	1	0.9968876	0	0.011280504
7157	0	0.31758755	0	0.10145683	1	0.94425654
9030	1	0.937251	1	0.14905281	1	0.17356524

Guns	Guns Prediction	Domestic Stress	Domestic Stress Prediction	Suicidality	Suicidality Prediction
1	0.633834	0	0.0061767236	1	0.9456499
0	0.0133809075	1	0.92579377	1	0.9422537
0	0.0052337176	0	0.012059539	0	0.022399604
0	0.010115084	0	0.03462818	0	0.10554715
0	0.009144158	1	0.106913686	0	0.17868902

FIGURE 7.22   Actual value vs predicted value with RNN using Keras preprocessing.

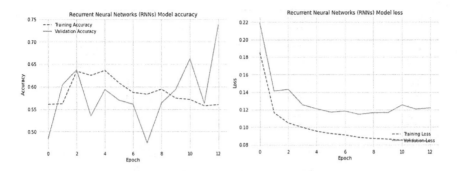

FIGURE 7.23 Graphs depicting training vs testing accuracy and loss over epoch for RNN using Keras preprocessing model.

Post ID	Economic Stress	Economic Stress Prediction	Isolation	Isolation Prediction	Substance Use	Substance Use Prediction
262	1	0.023568435	0	0.055551488	0	0.10392198
377	0	0.120327175	0	0.10997115	0	0.13218188
419	0	0.026378844	1	0.077478625	0	0.18332429
7157	0	0.20597932	0	0.20304005	1	0.9691302
9030	1	0.35537198	1	0.1854862	1	0.48466787

Guns	Guns Prediction	Domestic Stress	Domestic Stress Prediction	Suicidality	Suicidality Prediction
1	0.06562447	0	0.051457245	1	0.9843647
0	0.057882104	1	0.58095145	1	0.9860388
0	0.0060294084	0	0.058674738	0	0.18665063
0	0.023625094	0	0.13424276	0	0.16642101
0	0.0114129465	1	0.22518714	0	0.47074446

FIGURE 7.24 Actual value vs predicted value with RNN using GloVe.

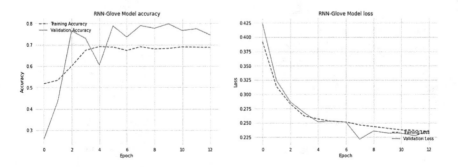

FIGURE 7.25 Graphs depicting training vs testing accuracy and loss over epoch for RNN-GloVe model.

Post ID	Economic Stress	Economic Stress Prediction	Isolation	Isolation Prediction	Substance Use	Substance Use Prediction
262	1	0.1623751	0	0.08528816	0	0.14324555
377	0	0.2080728	0	0.26798537	0	0.43574414
419	0	0.2803667	1	0.91977316	0	0.18756425
7157	0	0.8000599	0	0.2616184	1	0.98438346
9030	1	0.8870542	1	0.32123914	1	0.57614535

Guns	Guns Prediction	Domestic Stress	Domestic Stress Prediction	Suicidality	Suicidality Prediction
1	0.01762572	0	0.033553854	1	0.8193319
0	0.035806753	1	0.91459453	1	0.96223426
0	0.016501294	0	0.042056512	0	0.1002131
0	0.021555081	0	0.065508515	0	0.118149824
0	0.0067484146	1	0.0621125	0	0.10192474

FIGURE 7.26 Actual value vs predicted value with RNN using Word2Vec.

7.4.6.2.3 RNN Using Word2Vec
Figure 7.26 shows the result of implementing RNN using Word2Vec.
Figure 7.27 shows graphs depicting training vs testing accuracy and loss over epoch, respectively, for RNN-Word2Vec model.

7.4.6.2.4 RNN Using FastText
Figure 7.28 shows the result of implementing RNN using FastText
Figure 7.29 shows graphs depicting training vs testing accuracy and loss over epoch, respectively, for RNN-FastText model.

7.4.7 DISCUSSION

In this section, we summarise the accuracies of the neural networks based on the different embedding models that are used as shown in Table 7.3. We can observe the

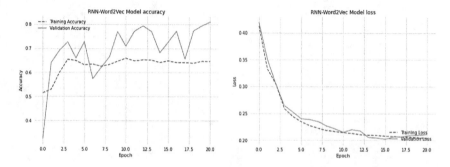

FIGURE 7.27 Graphs depicting training vs testing accuracy and loss over epoch, respectively, for RNN-Word2Vec model.

Post ID	Economic Stress	Economic Stress Prediction	Isolation	Isolation Prediction	Substance Use	Substance Use Prediction
262	1	0.13201067	0	0.06917658	0	0.11994362
377	0	0.18402387	0	0.08326335	0	0.11328109
419	0	0.17832512	1	0.07035875	0	0.11901687
7157	0	0.6069383	0	0.16982561	1	0.9999999
9030	1	0.6173112	1	0.12274662	1	0.94507384

Guns	Guns Prediction	Domestic Stress	Domestic Stress Prediction	Suicidality	Suicidality Prediction
1	0.011380536	0	0.07192046	1	0.9986778
0	0.010310583	1	0.069503985	1	0.99114275
0	0.004371463	0	0.06294445	0	0.07221062
0	0.00567246	0	0.22867464	0	0.19553606
0	0.00040603528	1	0.2017143	0	0.13915406

FIGURE 7.28   Actual value vs predicted value with RNN using FastText.

models' testing accuracy is highest with CNN with GloVe embedding and with RNN with FastText. The testing accuracy is lowest while the models were trained using Keras preprocessing libraries. The GloVe embedding provides higher test accuracies compared to other embedding models.

7.5   FUTURE SCOPE

In this chapter, we have analysed the reasons behind depression, by classifying them into five categories, economic difficulties, isolation, substance use, gun violence and domestic stress. In the future, we can map this work into a regression problem where the severity of each of these problems can be evaluated. With the help of machine vision and speech, an audiovisual dataset that includes videos of individuals may help to identify the causes of depression more accurately and find out their impact on the depressed person's life. Moreover, with the help of machine learning, we can

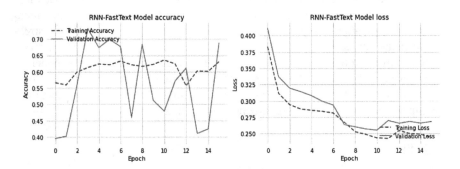

FIGURE 7.29   Graphs depicting training vs testing accuracy and loss over epoch, respectively, for RNN-FastText model.

TABLE 7.3
Summary of Training and Testing Accuracies

Model	Training Accuracy (%)	Testing Accuracy (%)
Convolutional neural network (CNN)	57.04	54.47
Recurrent neural networks (RNNs)	54.98	50.10
Convolutional neural network (CNN) with GloVe	56.03	73.92
Recurrent neural networks (RNNs) with GloVe	62.82	70.84
Convolutional neural networks (CNNs) with Word2Vec	64.68	68.09
Recurrent neural networks (RNNs) with Word2Vec	65.78	64.72
Convolutional neural networks (RNNs) with FastText	66.50	64.46
Recurrent neural networks (RNNs) with FastText	61.97	72.96

try to implement an online bot that can act as a psychologist, providing advice to mentally ill people. It must be kept in mind that introverted people may not be comfortable admitting that they are depressed and may not speak freely even to a doctor. So, opening up to a chatbot is easier for them. We have analysed texts from Reddit posts here, but by supervising a person's activity on social media like by analysing a playlist, the songs that person is playing frequently, or types of literature or movies he/she prefers, all of these can be used as metrics to detect the levels of mental health and take actions if necessary. In this way, we can help the persons in need.

7.6 CONCLUSION

This chapter has attempted to detect the possible reason behind depression from social media posts on sub-Reddit. The applicability of CNN and RNN each having four variations in embedding has been evaluated to classify the reason among economic difficulties, isolation, substance use, gun violence and domestic stress. Whether the concerned person has a suicidal tendency or not has been attempted to be identified from the post. The accuracy of each deep neural network model has been considered as the performance metric. Performance of each model has been compared with each other and in future can be attempted to improve and more models can be tested to have a future direction in this area of research.

REFERENCES

AlSagri, H.S. and Ykhlef, M., 2020. Machine learning-based approach for depression detection in twitter using content and activity features. *IEICE Transactions on Information and Systems*, *103*(8), pp. 1825–1832.

Antoni, M.H., Lehman, J.M., Kilbourn, K.M., Boyers, A.E., Culver, J.L., Alferi, S.M., Yount, S.E., McGregor, B.A., Arena, P.L., Harris, S.D. and Price, A.A., 2001. Cognitive-behavioral stress management intervention decreases the prevalence of depression and enhances benefit finding among women under treatment for early-stage breast cancer. *Health Psychology*, *20*(1), p. 20.

Biswas, S., Ghosh, A., Chakraborty, S., Roy, S. and Bose, R., 2020. Scope of sentiment analysis on news articles regarding stock market and GDP in struggling economic condition. *International Journal*, *8*(7), pp. 3594–3609.

Bojanowski, P., Grave, E., Joulin, A. and Mikolov, T., 2017. Enriching word vectors with subword information. *Transactions of the Association for Computational Linguistics*, *5*, pp. 135–146.

Cepoiu, M., McCusker, J., Cole, M.G., Sewitch, M., Belzile, E. and Ciampi, A., 2008. Recognition of depression by non-psychiatric physicians—a systematic literature review and meta-analysis. *Journal of General Internal Medicine*, *23*(1), pp. 25–36.

Cho, K., Van Merriënboer, B., Gulcehre, C., Bahdanau, D., Bougares, F., Schwenk, H. and Bengio, Y., 2014. Learning phrase representations using RNN encoder-decoder for statistical machine translation. *arXiv preprint arXiv:1406.1078*.

Church, K.W., 2017. Word2Vec. *Natural Language Engineering*, *23*(1), pp. 155–162.

Cohn, J.F., Kruez, T.S., Matthews, I., Yang, Y., Nguyen, M.H., Padilla, M.T., Zhou, F. and De la Torre, F., 2009, September. Detecting depression from facial actions and vocal prosody. In *2009 3rd international conference on affective computing and intelligent interaction and workshops* (pp. 1–7). IEEE.

Dadiz, B.G. and Ruiz, C.R., 2019. Detecting depression in videos using uniformed local binary pattern on facial features. In *Computational science and technology* (pp. 413–422). Springer, Singapore.

Guntuku, S.C., Yaden, D.B., Kern, M.L., Ungar, L.H. and Eichstaedt, J.C., 2017. Detecting depression and mental illness on social media: an integrative review. *Current Opinion in Behavioral Sciences*, *18*, pp. 43–49.

Islam, M.R., Kabir, M.A., Ahmed, A., Kamal, A.R.M., Wang, H. and Ulhaq, A., 2018. Depression detection from social network data using machine learning techniques. *Health Information Science and Systems*, *6*(1), pp. 1–12.

Ketkar, N., 2017. Introduction to keras. In *Deep learning with Python* (pp. 97–111). Apress, Berkeley, CA.

Klein, D.N. and Santiago, N.J., 2003. Dysthymia and chronic depression: introduction, classification, risk factors, and course. *Journal of Clinical Psychology*, *59*(8), pp. 807–816.

Koo, J., Marangell, L.B., Nakamura, M., Armstrong, A., Jeon, C., Bhutani, T. and Wu, J.J., 2017. Depression and suicidality in psoriasis: review of the literature including the cytokine theory of depression. *Journal of the European Academy of Dermatology and Venereology*, *31*(12), pp. 1999–2009.

Liu, H., Lang, B., Liu, M. and Yan, H., 2019. CNN and RNN based payload classification methods for attack detection. *Knowledge-Based Systems*, *163*, pp. 332–341.

Loades, M.E., Chatburn, E., Higson-Sweeney, N., Reynolds, S., Shafran, R., Brigden, A., Linney, C., McManus, M.N., Borwick, C. and Crawley, E., 2020. Rapid systematic review: the impact of social isolation and loneliness on the mental health of children and adolescents in the context of COVID-19. *Journal of the American Academy of Child & Adolescent Psychiatry*, *59*(11), pp. 1218–1239.

Lopez Molina, M.A., Jansen, K., Drews, C., Pinheiro, R., Silva, R. and Souza, L., 2014. Major depressive disorder symptoms in male and female young adults. *Psychology, Health & Medicine*, *19*(2), pp. 136–145.

Low, D.M., Rumker, L., Talker, T., Torous, J., Cecchi, G. and Ghosh, S.S., 2020. Reddit Mental Health Dataset. doi:10.17605/OSF.IO/7PEYQ

Mowery, D., Bryan, C. and Conway, M., 2017. Feature studies to inform the classification of depressive symptoms from Twitter data for population health. *arXiv preprint arXiv:1701.08229.*

Parameswaran, N.S. and Venkataraman, D., 2019. A computer vision based image processing system for depression detection among students for counseling. *Indonesian Journal of Electrical Engineering and Computer Science, 14*(1), pp. 503–512.

Pennington, J., Socher, R. and Manning, C.D., 2014. GloVe: global vectors for word representation.

Ramírez-Cifuentes, D. and Freire, A., 2018. UPF's participation at the CLEF eRisk 2018: early risk prediction on the Internet. In Cappellato L, Ferro N, Nie JY, Soulier L, editors. *Working notes of CLEF 2018 – Conference and labs of the evaluation forum; 2018 Sep 10-14; Avignon, France.[Avignon]: CEUR workshop proceedings; 2018* (pp. 1–12). CEUR Workshop Proceedings.

Saha, S. and Ghosh, A., 2019, December. Rehabilitation using neighbor-cluster based matching inducing artificial bee colony optimization. In *2019 IEEE 16th India council international conference (INDICON)* (pp. 1–4). IEEE.

Sellick, S.M. and Crooks, D.L., 1999. Depression and cancer: an appraisal of the literature for prevalence, detection, and practice guideline development for psychological interventions. *Psycho-Oncology: Journal of the Psychological, Social and Behavioral Dimensions of Cancer, 8*(4), pp. 315–333.

Sharma, S., Sharma, S. and Athaiya, A., 2017. *Activation functions in neural networks. towards data science, 6*(12), pp. 310–316.

Smith, M.E., Sharpe, T.L., Richardson, J., Pahwa, R., Smith, D. and DeVylder, J., 2020. The impact of exposure to gun violence fatality on mental health outcomes in four urban US settings. *Social Science & Medicine, 246*, p. 112587.

Stack, S.J., 2014. Mental illness and suicide. *The Wiley Blackwell Encyclopedia of Health, Illness, Behavior, and Society*, pp. 1618–1623.

Suthaharan, S., 2016. Decision tree learning. In *Machine Learning Models and Algorithms for Big Data Classification* (pp. 237–269). Springer, Boston, MA.

Swinger, N., De-Arteaga, M., Heffernan IV, N.T., Leiserson, M.D. and Kalai, A.T., 2019, January. What are the biases in my word embedding? In *Proceedings of the 2019 AAAI/ACM conference on AI, ethics, and society* (pp. 305–311).

Venkataraman, D. and Parameswaran, N.S., 2018. Extraction of facial features for depression detection among students. *International Journal of Pure and Applied Mathematics, 118*(7), pp. 455–463.

Vinchurkar, D.P. and Reshamwala, A., 2012. A review of intrusion detection system using neural network and machine learning. *Journal of Engineering Science and Innovative Technology, 1*, pp. 54–63.

Zhang, K., Zuo, W., Gu, S. and Zhang, L., 2017. Learning deep CNN denoiser prior for image restoration. In *Proceedings of the IEEE conference on computer vision and pattern recognition* (pp. 3929–3938).

8 Psychiatric Chatbot for COVID-19 Using Machine Learning Approaches

Priyanka Jain, Subhash Tatale, Nivedita Bhirud,
Sanket Sonje, Apurva Kirdatt, Mihir Gune and
N. K. Jain

CONTENTS

DOI: 10.1201/9781003122401-8

8.1   INTRODUCTION

After the outbreak of coronavirus disease 2019 (COVID-19), a number of businesses and industries suffered huge downfalls, millions of people lost their jobs and all these led to an increase in mental disorders and disturbances among people. Along with mental issues, the worldwide spread of COVID-19 has made people suffer from a number of misconceptions and misbeliefs. It became difficult to access accurate, official and verified information regarding the disease. After deep research and study of COVID-19, research institutes found that aftereffects of the disease are mentally tiring. Serafini et al. (2020) states that nearly one in five people who have had COVID-19 are diagnosed with a psychiatric disorder within 90 days, where anxiety, depression and insomnia were the most common mental ailments among recovered patients.

The number of mentally ill people in the world has increased manifoldly. Due to the increasingly remote way in which people are starting to lead their lives, it is safe to say that mental illness is a problem that is bound to increase. Lack of companionship, a feeling of discontent and anxiety, which may end up in substance addiction, are very common problems with the current millennial generation. Millennials are also more likely to confide their trust in an e-counselor. To address these important issues faced by society in recent times, the combined chatbot system is proposed, which mainly revolves around two research domains, machine learning (ML) and natural language processing (NLP), and helps tackle both the issues simultaneously.

ML and NLP are intensive techniques in which we can train machines as per our requirement. Chatbots, which are one of the most popular examples of ML, are becoming more and more popular in fields such as business, research, education and, most importantly, medical industry (Sancheti et al. 2020). Today, chatbots have replaced experienced physicians and doctors and are taking over a wide range of applications from minor disease diagnosis to addressing mental issues with the same efficiency and accuracy (Rahman et al. 2017). The study done by Bhirud et al. (2019) suggests that chatbots are as capable as humans in communication and can simultaneously hold conversations while producing real-time output.

On the other hand, NLP enables the system to understand the conversational language spoken by humans. The conversational language employed by humans for day-to-day conversations isn't as perfect as the formal language. This language does not focus much on vocabulary and grammar. Most languages follow different grammatical rules in spoken and written forms. Hence, it is very difficult for the system to deal with the language used by the user. The system receives input from the user in the form of natural language. This is in unstructured text format, which cannot be understood by the system. The system understands input only in structured formats. To convert unstructured data to structured data, some steps are used for conversion.

We believe that tools, such as our chatbot, have a massive scope and would play an effective part in dealing with young people's mental health problems. The chatbot will be capable of answering questions related to disease and predicting the chances of contracting COVID-19 very precisely. It will also be able to constantly monitor the user's depression levels through regular Patient Health Questionnaire-9 (PHQ-9) (Richardson et al., 2010) tests and also keep tabs on topics and problems in the user's life. This can be converted into a very useful dataset that the user's psychologist can refer to during a face-to-face meeting.

In the future, it can be expanded to other diseases in which sufficient data connecting symptoms to contraction is available. In addition, our current system is only able to converse with the user in English; thus, it expects rudimentary English proficiency.

Section 8.2 explores on survey of mental health chatbot. Section 8.3 presents the need, motivation and the challenges of proposed system. Section 8.4 describes architecture and methods. In Section 8.5, database and resources are explained. Results and findings are discussed in Section 8.6. Section 8.7 concludes the chapter by discussing some further aspects.

8.2   LITERATURE SURVEY

A number of algorithms and ML models can be used to build smart and more advanced chatbots. Chatbots are being used for different purposes and can be implemented with various approaches. The intention behind this survey was to become familiar with possible ML algorithms and NLP concepts that can be used to make the system more interactive and precise. A simple chatbot system consists of modules such as natural language understanding (NLU), NLP, natural language generation (NLG), chat interface, a database and decision engine.

In the following section, reviews on Chat Interface, Word Tokenisation, COVID-19 Remote Tests and Seq2Seq Transformer Models are presented. Furthermore, our Literature Survey will continue with reviews on Mental Health Analysis, Sentiment Analysis and PHQ-9 test.

8.2.1   Chat Interface

The chat interface of a chatbot is the entire front end of the chatbot. When a query is inserted into the chat interface, it is processed at the back end and then the appropriate response is shown in the chat interface. A good chatbot interface can be designed using popular User Experience (UX) Design guidelines and research on various human-computer interaction (HCI) tools available.

An effective chatbot interface should allow rapid and decisive interaction with the consumer. The technologies used for building a chat interface depend upon the environment in which the chatbot will be located. Several existing frameworks can be used to create the chatbot interface (e.g., BotUI is a JavaScript framework). Chatbots can also be hosted on existing ML platforms, including WhatsApp and Facebook Messenger.

It can be deployed in various forms as per requirements of the application and users who are going to use it. It can either be in the form of an android app or as a web application. In the case of an android app, the user query is to be typed into the Textbox like a normal message on a messaging app and chatbot response will be given in the same way. In web-based bot, the visiting user will be advised to chat with the Web Bot in the case of any queries. The rest of the process remains the same for both methods.

8.2.2 Word Tokenisation

The process of mapping a word from vocabulary to an integer value so that it can be used in the further process is called word embedding. Some of the techniques are mentioned in below section:

Python tokeniser is available in the Python NLTK library, which has various functions and is built to help us with the entire NLP methodology. NLTK stands for Natural Language Toolkit and is basically used for the conversion of machine language into human readable or natural language. It is a powerful Python NLP library most widely used for processing of text such as stemming, tokenisation and word count. Using this tokeniser and the vocabulary, we can convert a string into a sequence of IDs, i.e., integers. Each word in the vocabulary has to be given specific IDs so that they can easily be understood by the machine.

8.2.3 NLP Engine

NLP is a process of recognising the natural language input from the user, structuring, cleaning, preprocessing and making it ready for further processing. There are two subparts of processing the input, data cleaning and preprocessing.

Data cleaning and preprocessing are some of the most important processes of NLP. It includes removing symbols, grammatical errors, duplicate values, typos and missing and irrelevant data from the dataset. Data de-duplication and validation are one of the major tasks of this stage. By assuring that the data is error-free and grammatically correct, it is validated and made available for the data preprocessing. This is the stage where the special characters and symbols from the data are removed and data is lowered. After that, to understand the sentences and break them down, a word tokeniser is used. Various data cleaning and software tools are used to maintain the above-mentioned qualities of data making the process even smoother. Also, we can write a small function to clean the data and preprocess it.

8.2.4 COVID-19 Remote Test

COVID-19 testing advisory system is considered to build proposed chatbot. This can be brought about by checking the user for all possible symptoms and then using an ML model with those symptoms as its inputs. The model is used to calculate the probability of the user being infected with COVID-19 (Menni et al., 2020). During the research, it is apparent that as COVID-19 is a new disease, there is a lack of research materials with respect to predicting the likelihood of individuals contracting it.

Sisodia and Sisodia (2018) in a relevant paper have used classification algorithms to predict disease likelihood. They have used three classification algorithms – support vector machines (SVMs), naïve Bayes classifier and a decision tree classifier – to determine the likelihood of diabetes in patients. As an input to the classifier, several attributes such as diastolic blood pressure, skin fold thickness and BMI are used. All three algorithms are trained using a dataset containing all the attributes listed previously and the information whether the respective person has diabetes or not (i.e., supervised learning). After the prediction on a large scale, it is concluded that the naïve Bayes classifier outperforms the other two algorithms.

Others have used three classification algorithms (Thirugnanam, 2013) – SVMs, decision trees and logistic regression. The previous algorithms are used to determine the likelihood of the patient contracting a cardiovascular disease. Like the above-mentioned example, they have used 13 attributes; some of them are age, sex, chest pain and cholesterol.

8.2.5 Seq2Seq Transformer Model

A general sequence-to-sequence (Seq2Seq) model is a neural network that converts the given input sequence into another sequence. The word transformer here adds an attention mechanism to encoder-decoder models. A Seq2Seq model takes an input sequence, converts it and maps it to a vector, which, in turn, goes as an input to the decoder, and respective output is generated.

In the case of the transformer Seq2Seq model, attention mechanism checks the input sequence and deduces the essential components of the sequence (Varshney et al., 2020). It considers all possible inputs and, at the same time, assigns them weights. This encoded sequence goes as an input to the decoder. The decoder learns the word sequence that comes from encoder output and predicts the sentences based on all the words before the current word and, based on this, the final output will be generated. Recurrent neural network (RNN) doesn't see any hierarchy in a sequence. The models alter the hidden state every time a new input is being processed despite how insignificant it might be. Therefore, the data from earlier layers may end up being completely washed out by the time the RNN reaches the end of the sequence. While, in the transformer model, the output has a comparatively better context because there is no memory issue.

Another inherent disadvantage of RNN or long short-term memory (LSTM) lies in the nature of sequential processing. Since parts of the input are processed one at a time (H2 cannot be computed without H1), the networks' computations, overall, are very slow. The transformer can compute output in parallel, for all steps at once. Improving models against adversarial attacks is an important field of research in ML. It appears to be that transformer models are more robust against adversarial attacks.

Considering all the previous points, the Seq2Seq transformer model plays an important role in language translation as well as language generation tasks. Hence, the authors used the Seq2Seq transformer model over other neural network models such as LSTM, RNN.

8.2.6 Mental Health Analysis

The idea of using chatbots for mental health problems dawned in 1972 when PARRY, a program that could mimic the behavior of a human with schizophrenia, was developed. During the last several years, due to the development of better and better conversational algorithms, several mental health chatbots have been in use. Confidentiality is still one of the major concerns but a lot of research is being done in this field. There are chatbots that are being targeted toward specific mental illnesses too.

It has been inferred (Ayanouz et al., 2020) that one of the most famous psychiatric chatbots, called Casper (or Insomnobot-3000), is the only bot in the world that is available for a chat between 12 pm and 4 am for people suffering from insomnia.

Woebot, a chatbot designed by Woebot Labs in San Francisco (Fitzpatrick et al., 2017), is a chatbot that treats depression and anxiety. Chatbots are also being used to target specific populations, such as HARR-E, which is a chatbot that fights against male suicide.

Due to the increasing success of mental health chatbots, some have predicted that chatbots may be the future of mental health and may end up replacing mental health professionals altogether. However, it is still not possible for chatbots to provide professional mental health advice to patients. Chatbots are not and will never be equivalent to a psychiatrist.

8.2.6.1   Sentiment Analysis

The purpose of a psychiatric chatbot is, in the simplest of terms, to act as a pocket psychiatrist for the user. Human psychiatrists react to their patients' moods and emotions and help them get through a difficult time and put their thoughts in order. In order to play this role of a pocket psychiatrist, our bot needs to know the emotional status of the user. Thus, an ML algorithm is designed to gauge user's feelings and emotions. Following are a few of the ways in which emotion recognition is carried out.

A GRU-based RNN can be used (Lee et al., 2017) to extract the utterance intention from the query. This neural network is used to become familiar the intention of the document and extracting entities. Additionally, other user information, such as location, gender, and sex, are collected through wearable devices. Thus, we must note that a closer and more continuous inspection of the user's mental health is carried out before emotion recognition. The output from the GRU is represented in a vector, which represents the extracted features of the text. Then, another neural network is used to decide the response of the chatbot depending on the intention features detected. This neural network determines the intention features of the text. These features are passed onto a response-generating neural network, which elicits a sympathetic response depending on the intention features.

A Bi-LSTM-based sentiment analysis model can also be deployed (Yin et al., 2019). In addition to the sentiment analysis model, the architecture of their system also has two chatbots, one is a normal conversational chatbot and the other is a psychiatric conversation chatbot. The purpose of the Bi-LSTM model is to predict the given text into the positive and negative text. This is done by fitting a sigmoid function to the output of the model. Any value less than 0.5 is a negative value and any value greater than 0.5 is a positive value. This sentiment analysis model is responsible for detecting the mood of the user for the entire conversation. A classifier model of identical architecture to the sentiment analysis model is also used. This model is used to filter the dataset into a psychological and an unrelated datasets. Thus, using this model, the psychological chatbot can be trained on the psychological dataset only.

8.2.6.2   PHQ-9 Test

There are several ML- and NLP-based solutions that can be used to understand the mood of a person through their speech. The TextBlob library is one such solution that deployed for sentiment analysis in the system. It is very effective in finding whether a particular statement is a "positive statement," "negative statement" or a "neutral statement." This has been used in our system to gauge the mood of the user and if found to be showing depressive indications, the user is then asked to submit to a detailed depression analysis test, explained in the following.

The PHQ-9 (Richardson et al., 2010) is a medically approved test used to detect the level of depression of an individual. In the test, the person is asked to quantify the number of times they felt a particular feeling over a specific time period. Each question has four options and each option has further specific points that are given on the basis of what number of times they felt a particular feeling. The test is self-administered.

8.3   NEED, MOTIVATION AND CHALLENGES

The COVID-19 pandemic has turned the lives of many people upside down. The lockdowns imposed to curb the spread of the virus have led to a major increase in the number of mentally ill people. An estimated 12.2 crore Indians lost their jobs in the month of April, according to the Centre for Monitoring Indian Economy. In addition to loss of income, long-term separation from their loved ones created a feeling of loneliness and contributed to the mental health crisis. As of November 2020, more than one lakh Indians have lost their lives due to the coronavirus. Bereaved families have found themselves unable to cope with the grief of losing their loved ones so suddenly. As a result, we currently face the largest mental health crisis in Indian history. The Indian Psychiatric Society estimates that the number of mentally ill Indians rose by 20% during the initial lockdown period.

Despite the fact that we live in the information age, it has been noticed that a large proportion of people do not have accurate and verified information about the coronavirus and the ensuing COVID-19 pandemic. This includes basic precautions, relevant emergency procedures and government rules and regulations. Due to unofficial and unverified sources, rumors and conspiracy theories have been rampant. Bad health information may lead to people disregarding the severity of the pandemic and thus endangering themselves as well as others. Online rumors targeting specific communities have even caused riots. Thus, it is extremely important that every Indian has access to accurate and verified information.

The developed chatbot is geared toward solving the problems described previously. First, the mental health chatbot (Jwala et al. 2019) interacts with the user in the same way as a psychiatrist would do with their patients. It acts as a counselor to the user. Additionally, it is capable of determining whether the user exhibits depressive symptoms, and then through the PHQ-9, it is able to deduce the level of depression the user is suffering from. Our chatbot is also a medium to provide the user with accurate information about the pandemic and answer any questions they may have about the crisis. Our COVID-19 tester module is able to provide the user with their probability of contracting COVID-19 by analysing their symptoms.

According to reliable estimates, only 10% of the Indian population speak English. Thus, NLPs capable of recognising Indian languages have huge potential. The challenges are numerous and diverse. First, Indian languages are ambiguous. Often, it can be hard to find accurate grammatical rules. In addition, there are disparities between speaking a language and its grammatical rules. In addition, the massive diversity within a language is also hard to deal with. Hindi, in itself, has seven major dialects, which individually have millions of speakers. So, an NLP system may need to be trained separately for each such dialect. In conjunction with this, finding appropriate conversational data to train such systems in each language and each dialect of that language is the hardest task of all.

Semantic analysis is the usage of algorithmic techniques to enable machines to recognise the context behind words and sentences. The biggest hindrance to semantic analysis is that it often requires long-term memory to correctly recognise the context. The algorithm may have to look at three or four previous sentences to obtain the correct answer.

Obtaining accurate and genuine mental health data to train a mental health and psychiatric analysis algorithm is extremely challenging. First, as mental health is a very private and personal topic, it is extremely rare that a conversation between a mental health patient and a professional is available in the open domain. In addition, mental problems vary a lot from person to person. The mental health data from an American source would not be appropriate to train a system that would be used by Indian people. The data will be enormously biased toward American Mental Health and be unusable in an Indian context. Another problem is that mental health conversations are extremely contextual. This context depends on the patient's own problems and thus this kind of data cannot be used to train algorithms.

The RNN-based Seq2Seq model has been the mainstay of most modern NLP chatbots. However, these models cannot still yet deal with long-range dependencies. The input of such a model is sequential. This prevents parallelism.

8.4   ARCHITECTURE AND METHODS

The detailed architecture is described in the following sections. Section 8.4.1.1 describes the COVID-19 Prediction Engine. Section 8.4.1.2 discusses the Conversational Module and its working. Sentiment analysis, NLP and ML model are the three major submodules of the Conversational module. Section 8.4.1.3 explains our Graphical User Interface.

8.4.1   System Architecture

In order to tackle the problem (described in Section 8.3), the authors proposed a system that has an app as an interface. The system has two main modules. One of them is COVID-19 Prediction Engine and the other is the Conversational Module. The project exploits the capabilities of ML to build the transformer algorithm. Figure 8.1 illustrates the entire architecture of the system. In the next sections, we will see each module in detail.

The first step for the user is to Log In/Sign Up to the system and for the very first time, the user has to give the COVID-19 test. After that, if the user wishes, the user can talk to the Co-Bot to get answers to queries related to COVID-19 and mental

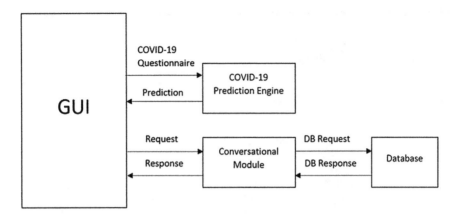

FIGURE 8.1 Architecture of the proposed system.

health. When the user finishes their conversation, the user receives their sentiment analysis report and if found depressed then we encourage the user to take the PHQ-9 test that is a worldwide accepted test and promoted by the World Health Organisation (WHO); otherwise, the user can exit from the window. The flow diagram of the proposed system is shown in Figure 8.2.

8.4.1.1 COVID-19 Prediction Engine

COVID-19 testing advisory system is used to build the proposed chatbot system. This can be brought about by checking the user for all possible symptoms and then using the COVID-19 formula with those symptoms as its input. The COVID-19 contraction formula is mentioned in Table 8.1. The model is used to determine the likelihood of the user being infected with COVID-19. The formula used to predict this probability can be seen in the following. This formula can only predict if the user is contracting COVID-19 or not and fails to predict whether the user has starting symptoms of COVID-19 or not. Hence, the authors proposed a new feature to this module in order to make it more accurate to predict the likelihood of contraction.

$$
\begin{aligned}
\text{Prediction Model} = {} & -1.32 - (0.01 * \text{age}) \\
& + (0.44 * \text{sex}) \\
& + (1.75 * \text{loss of smell and taste}) \\
& + (0.31 * \text{severe or significant persistent cough}) \\
& + (0.49 * \text{severe fatigue}) \\
& + (0.39 * \text{skipped meals}) \quad\quad\quad\quad (8.1)
\end{aligned}
$$

A one-dimensional array of size equal to the number of symptoms is used, i.e., four (loss of smell and taste, cough, fatigue and loss of appetite), which initialises all the symptoms to zero. At the time of the test, we ask questions to the user about

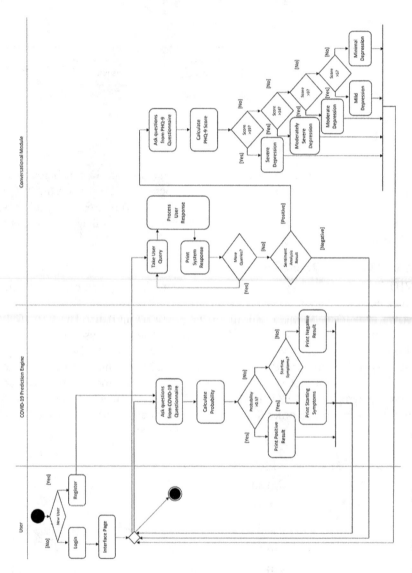

FIGURE 8.2   Flow diagram of the proposed system.

TABLE 8.1
Summary of the Literature Survey

Sr. No.	Authors	Problem Discussed and Solved	Method/Algorithm/Tools Used	Result
1	Sisodia and Sisodia (2018)	Predicting the likelihood of disease using classification algorithms	Decision tree classifier, support vector machine (SVM) and naive Bayes classifier	Three algorithms are used to compute the likelihood of contracting a disease
2	Lee et al. (2017)	An algorithm to recognise emotions from a query	RNN (GRU) used to classify the emotions into one of eight predefined stages	The GRU is used to break the sentence down into a vector
3	Su et al. (2017)	To increase the model accuracy	They suggested a two-layer LSTM model to deal with the semantic relation between words and sentences	Provides high accuracy, but more topics in dialog need to be provided and emotional and personality factors can be considered
4	Yin et al. (2019)	Detecting emotions in the user query and responding to them accordingly	Creating two chatbots (one casual and one psychological). Also using a Bi-LSTM-based sentiment analysis to determine user sentiment	The usage of multiple chatbots is unique. However, because corpus has been collected firsthand, bots are effective
5	Yusof and Man (2017)	How to store all the data to train the basic model?	Using JSON files, data can be stored and retrieved	JSON is very efficient and fast in retrieving data
6	Javed et al. (2015)	To implement word segmentation (tokenisation)	Calculating all character spaces	It involves mathematical calculations and hence proves to be slower than others
7	Vinyals and Le (2015)	An algorithm to generate a response to a query from scratch	An RNN-based Seq2Seq model (sequence-to-sequence model)	The model provides coherent responses despite its simplicity and a noisy dataset

(Continued)

TABLE 8.1 (*Continued*)
Summary of the Literature Survey

Sr. No.	Authors	Problem Discussed and Solved	Method/Algorithm/Tools Used	Result
8	Kataria et al. (2018)	An algorithm to generate a response from scratch but work around the vanishing gradient problem	An LSTM-based Seq2Seq model	The model provides much better recall over a larger time period than RNN-based models
9	Mathew et al. (2019)	A chatbot that encourages users to express their health-related doubts and concerns	K-nearest neighbor algorithm (KNN), Bag of Words (BoW)	A reliable system that can ascertain that the user is healthy by keeping tabs and thus helps in efficiently diagnosing diseases
10	Thomas and Thomas (2018)	Scaling up the chatbot to real-world domains with scripted dialogs	Gated end-to-end memory network, goal-driven dialog system	Access control for an end-to-end trainable memory-enhanced neural network
11	Sojasingarayar (2020)	Sometimes response is not relative to the question	Using probability, it is checked that how response is related to input. If probability is less than the threshold value, then the Seq2Seq model is used to auto-generate the response	An efficient and better response as compared to the previous method
12	Ling et al. (2015)	Part-of-speech tagging and dependency parsing using word2vec models	Structured Bag of Words (BoW), skip-gram model, continuous window model	Improvements in part-of-speech tagging and dependency parsing using modified versions of word2vec models

symptoms, if a particular symptom is present, then we just put 1 at the index that represents that symptom in the array. After asking all the questions, based on the formula, we calculate the probability of contracting COVID-19. If the probability is less than 0.5 and the number of ones in that array is zero, then the user is not contracting COVID-19. If the probability is less than 0.5 and the number of ones in that array is greater than zero, then the user might have starting symptoms of COVID-19. The last case is that if the probability is greater than 0.5, in that case, the user absolutely has COVID-19 symptoms and must contact the COVID-19 center nearby. This feature handles almost all the outliers and provides the best possible output.

8.4.1.2 Conversational Module

The Conversational Module is the heart of chatbot system. This module is responsible for doing several tasks mentioned in Figure 8.3.

The first task of the Conversational Module is sentiment analysis. In sentiment analysis, the user queries are analysed to detect whether the user query is depressive or neutral or positive. After sentiment analysis, a query is passed to the NLP module. In this process, mainly focus is given on cleaning user queries, preprocessing it to make it ready for further processing. The ML model is used to predict the response to the input query. When the user stops asking questions then the system displays

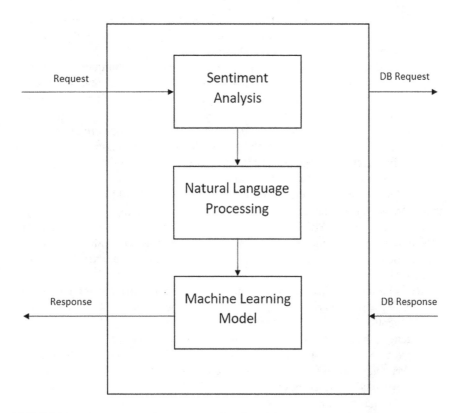

FIGURE 8.3   Components of the conversational module.

the result of sentiment analysis and advises the user about taking the PHQ-9 test. This is how the Conversational Module works and the detailed explanation of each submodule is given in the following.

Using sentiment analysis, it is inferred whether the text is positive, neutral or negative. This can be done by using the Python library "TextBlob." The main aim of sentiment analysis is to count the number of negative, neutral and positive statements. Initially, the counts of these variables are initialised to zero. When a user inputs a query to the chatbot, the system performs sentiment analysis on that query and increment the count of the respective variable. When a user exits from the chatbot window or stops asking questions at that time, the counts of negative, positive and neutral statements are displayed. If the count of negative statements is higher than the positive and neutral statements, then it is concluded that the particular user may have depression and the system encourages the user to give the PHQ-9 test that has been developed to analyse the depression level of users and is accepted worldwide. If the count of positive or neutral statements is higher than negative statements, then the option is given to the user for the PHQ-9 test. The user will be asked to fill a questionnaire. The PHQ-9 tester will calculate the user's PHQ-9 score according to the answers the user provides. After the questionnaire is completed, the final depression score of the PHQ-9 test is calculated and displays the depression level.

NLP is one of the technologies that play an important role in chatbot development and implementation. It converts the natural or human-readable language into machine readable. In short, it makes the language used in our day-to-day life understandable to the machines using some processing techniques. The natural language is not in a proper structure and hence referred to as unstructured, but the machine only understands structured language and hence there is a need for conversion.

To convert unstructured data to structured data, some steps are used for conversion. The first step is to convert the user query into sentences. Second, the special characters are removed and the data is lowered for further uses. Also, the concatenated words such as "don't" and "we'll" are separated and converted to "do not" and "we will," respectively, because when we build our vocabulary, at that time, we have a large number of common words and the size of the vocabulary can be reduced. After that, to understand the sentences, a word tokeniser is used to convert the sentences into words and assign a unique index to each word in the vocabulary. In the NLP unit, each word in the vocabulary is stored for further use.

ML is an intensive technique in which machines are trained as per our requirement. Using ML, a variety of operations can be performed and one of the most popular applications of ML is a chatbot.

A general Seq2Seq model is a neural network that converts the given input sequence into another sequence (Sutskever et al. (2014)). The word transformer here adds some attention mechanism to encoder-decoder models. In the Seq2Seq transformer model, the attention mechanism checks the input sequence and decides which other parts of the sequence are important. It considers all possible inputs at the same time and assigns them weights. The encoder of the transformer will clean the input text then encode the sequence by adding a pad sequence of fixed length. This encoded sequence goes as an input to the decoder. A decoder learns the word sequence that comes from encoder output and assigns weights as per importance and returns the linear layer. This linear layer has a sequence, which learns the embedding of words with respect to their importance in the sentence. The transformer takes the input query, again cleans and encodes the query and then predicts the words, one word at a time based on all the words before the current word and, based on this, the final output will be generated.

Several Seq2Seq transformers consist of the encoder-decoder system followed by some recurrent layers or a convolution layer. Transformers primarily focus on the attention mechanism, and hence instead of using the recurrent layer mechanism, they use multi-head attention that consists of multiple scalar dot product attention. Scalar dot product attention has three inputs: query, key and value. The scaled dot product attention is used to calculate the attention weights. The attention weight is nothing but how much importance factor is given to the particular query. To calculate the attention weights, the formula mentioned in Equation (8.2) is used:

$$\text{Attention}(Q, K, V) = \text{softmax}_k\left(\left(Q * K^T\right) / d_k\right) * V \qquad (8.2)$$

The SoftMax function is used in order to focus on the data at a particular instant and give less importance to irrelevant words for that particular query. Using this function, the weights are assigned to the given sequence. The transformer uses stacked multi-head attention layers and dense layers of both encoder and decoder. As we know that the transformer does not contain any convolution or recurrent layer, the positional encoding (PE) method is used in order to give the model some information about how the words are related in the sentence. For PE, the formula mentioned in Equation (8.3) is used:

$$PE_{(pos,2i)} = \sin\left(pos / 10000^{(2*i/D)}\right)$$
$$PE_{(pos,2i)} = \cos\left(pos / 10000^{(2*i/D)}\right)$$
$$\text{where } D = d_{model} \qquad (8.3)$$

Encoding layers have multi-head attention with a padding mask and two dense layers followed by dropout. Each encoder has three components, input embedding, two encoder layers and positional embedding. The encoder was plotted with the plot-model function in Keras. This encoder takes the input sequence and converts the input sequence into a continuous sequence of fixed length by using the pad sequencing method. Figure 8.4 implies the architecture layers of encoder of the model that converts the input sequence into a continuous sequence of fixed length by using the pad sequencing method.

Decoder layers masked multi-head attention with a look-ahead mask and padding mask, multi-head attention with a padding mask and two dense layers followed by

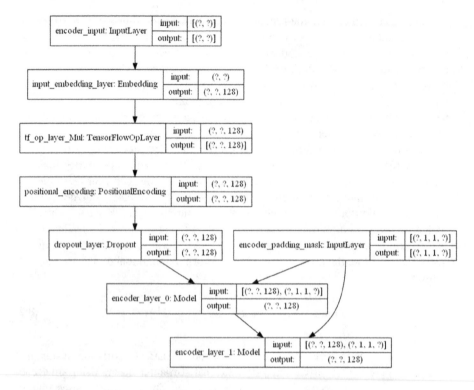

FIGURE 8.4 Architecture layers of the encoder.

dropout. Each decoder consists of output embedding, PE and two decoder layers. Figure 8.5 is the architecture diagram of the second important part of ML model, i.e., decoder. The output previously generated by the encoder is given as an input to decoder and with the understanding of word embedding, regaining the meaning and order of the sequence, it produces another sequence as an output. The architecture diagram of an ML model is shown in Figure 8.5.

The continuous sequence that is the output of the encoder is now passed to the decoder as an input and then the decoder understands the embedding sequence, how the words are placed together to form a sentence, which words are really important in the sentence, with the help of PE, help generate the output sequence. Figure 8.6 depicts the architecture of the transformer that tells how the layers of encoder, decoder and transformer are related to each other, how they are arranged to perform every small task in an efficient and better manner.

The Seq2Seq transformer algorithm is used to make predictions to the inputted user query as the main ML algorithm in the entire system.

8.4.1.3 Graphical User Interface

Our system has been converted into a productised form by integrating it into an android app. Thanks to the widespread usage and high penetration of the android

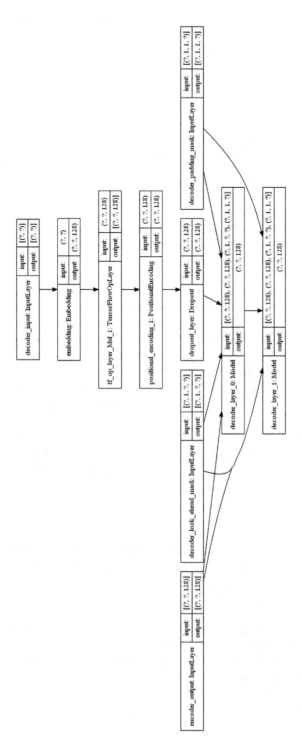

FIGURE 8.5 Architecture of the machine learning model.

```
Model: "transformer"
_____
Layer (type)                   Output Shape        Param #    Connected to
==================================================================================
inputs (InputLayer)            [(None, None)]       0
_____
dec_inputs (InputLayer)        [(None, None)]       0
_____
enc_padding_mask (Lambda)      (None, 1, 1, None)   0          inputs[0][0]
_____
encoder (Model)                (None, None, 512)    6312960    inputs[0][0]
                                                               enc_padding_mask[0][0]
_____
dec_look_ahead_mask (Lambda)   (None, 1, None, None 0          dec_inputs[0][0]
_____
dec_padding_mask (Lambda)      (None, 1, 1, None)   0          inputs[0][0]
_____
decoder (Model)                (None, None, 512)    8416256    dec_inputs[0][0]
                                                               encoder[1][0]
                                                               dec_look_ahead_mask[0][0]
                                                               dec_padding_mask[0][0]
_____
outputs (Dense)                (None, None, 6166)   3163158    decoder[1][0]
==================================================================================
Total params: 17,892,374
Trainable params: 17,892,374
Non-trainable params: 0
```

FIGURE 8.6 Architecture of the transformer model.

OS, the app will help us get our chatbot to reach the largest possible audience. Due to the high usage of mobile phones, making our chatbot usable on a smartphone significantly increases its usability and reach.

Our android app uses a Text Messenger such as UI to converse with the chatbot. The user query is to be typed into the Textbox like a normal message on a Messaging App. Meanwhile, when the ML model is predicting the response to the user query, the chatbot response will be shown in a bubble, in a similar fashion to common Messaging Apps. When the process of prediction is done then the response is shown in the Textbox, which is generated next to the input Textbox.

8.5 RESOURCES

Our database or dataset is broadly classified into two sections. The first section is the Mental Health Database and the second section is the COVID-19 Database.

8.5.1 MENTAL HEALTH DATABASE

The Mental Health Database consists of a set of questions that a user might ask the chatbot while having conversations. All these questions are collected from real-time question-answer sessions between a psychiatrist and a patient. The contents of the database are stored and accessed in a JSON format. The question-answer dataset is divided into separate lists of questions and answers for data preprocessing purposes and is stored separately. Figure 8.7 shows the question-answer format of the mental health dataset.

	Questions	Answers
0	A friend of my child's father is sending him i...	It sounds like you are wanting to protect your...
1	About a year ago I found out my husband had ch...	First of all, my heart goes out to you. Infide...
2	All I can do is cry and hate myself	Crying due to a dissolution of a marriage, is ...
3	Am I a lesbian because I love a girl?	It all depends how old you are and what a part...
4	Am I being picky when it comes to my boyfriend?	Well, then your decision is whether you want t...

FIGURE 8.7   Question-answer format of the Mental Health Dataset.

8.5.2   COVID-19 DATASET

The COVID-19 dataset consists of all COVID-19–related questions and their respective answers in the way as that of the psychiatric dataset. It contains all the COVID-19–related information that will help the user to get rid of doubts and misunderstandings about the disease. JSON objects are used as the basic data format. Figure 8.8 shows the JSON format of the COVID-19 dataset.

Fields in the JSON format of the COVID-19 dataset:

- Paragraphs – Contains all tags of questions
- Question lists – Contains all questions of specific tag
- Question – Contains question of specific tag
- Questions id – Contains an id for every question
- Answer's list – Contains all answers to the specific questions
- Answer – Contains the answer
- Answer id – Contains answer id

Then the JSON format data is converted into a question-answer data frame. Figure 8.9 shows the question-answer format of the COVID-19 dataset.

Final corpus is built by merging the mental health dataset and COVID-19 dataset, and for merging the datasets, Panda's library is used. Table 8.2 shows the statistic of the final corpus.

```
{"data": [
       {"paragraphs": [
              {"question_lists": [
                     {"question": "",
                      "questions_id":
                      "answers_list": [
                              {"answer":" ",
                               "answer_id":}]
                      }]
              }]
       ]}
```

FIGURE 8.8   JSON format of COVID-19 Dataset.

```
# check the questions and answers
df.head()
```

	Questions	Answers
0	What is the main cause of HIV-1 infection in c...	Mother-to-child transmission (MTCT) is the mai...
1	What plays the crucial role in the Mother to C...	DC-SIGNR plays a crucial role in MTCT of HIV-1...
2	How many children were infected by HIV-1 in 20...	more than 400,000 children were infected world...
3	What is the role of C-C Motif Chemokine Ligand...	High copy numbers of CCL3L1, a potent HIV-1 su...
4	What is DC-GENR and where is it expressed?	Dendritic cell-specific ICAM-grabbing non-inte...

FIGURE 8.9 Question-answer format of the COVID-19 dataset.

8.6 RESULTS AND FINDINGS

Using the ML and NLP techniques that are described in the previous sections, the authors developed a chatbot capable of interacting with the user in the same way a psychiatrist would do to their patient.

This section presents the results of a system along with the outcome. The system communicates with the user in the English language. The system uses natural conversational language in order to make the user comfortable. The system also provides the COVID-19 remote test that is completely free to everyone. The system provides the answers to related questions of the COVID-19 and mental health illness asked by the user. The system predicts the depression level of the user by using sentiment analysis followed by a PHQ-9 test.

8.6.1 EVALUATION RESULTS OF ML ALGORITHMS IN TERMS OF % ACCURACY

The authors trained the proposed model twice with different hyper-parameters. When training the model for the first time, the authors considered the number of layers $N = 2$, the number of heads $h = 8$, $d_{model} = 512$ and the dropout rate $P_{drop} = 0.2$ instead of 0.3, 300 training steps. For these inputs, the system gives a total of 58.73% of accuracy and 0.69% of loss in the model. This model is basically trained on an Intel i5 8th generation laptop having an 8-GB RAM and took around 3.13 hours to train. The authors observed that, after around 180–185 iterations (epochs), there is little change in model accuracy.

TABLE 8.2
Statistics of Final Corpus

	Total Number of Questions	Total Number of Answers
COVID-19 Dataset	514	514
Mental health Dataset	817	817
Total	1331	1331

TABLE 8.3
Accuracy of the Model

| | Hyper-parameters | | | | | | Training Accuracy | |
Sr. No.	Number of Layers (N)	d_model (d_model)	Number of Heads (h)	Units	Dropout	Steps	(%)	Loss (%)
1	2	512	8	512	0.2	300	58.73	. 0.69
2	2	256	8	512	0.1	300	58.96	0.62

While training the model for the second time on the same machine, the authors considered the number of layers N = 2, the number of heads h = 8, d_{model} = 256 and the dropout rate P_{drop} = 0.1 instead of 0.2, 300 training steps. For these inputs, the system gives a total of 58.96% of accuracy and 0.62% of loss in the model. The authors found no such big difference in the accuracy of the model, but it is suggested to change the parameters according to the need and size of the dataset. Table 8.3 gives more insights about the hyper-parameters.

8.7 DISCUSSION

This section discusses the contribution of the authors toward proposed system, conclusion of the chapter and future scope of the proposed system.

8.7.1 CONTRIBUTION

In today's world of fast-growing technology and smart communication, many chatbots are being implemented for various purposes in various domains. We can see a large number of chatbots in the marketing and corporate domain. Nowadays, chatbots are also being developed in the health-care domain to increase the accuracy of operations, to provide solutions at any point in time, and these chatbots have been helpful to both doctors and patients. Some chatbots are built specially to solve the problem of the user, give an appropriate solution to them and recommend the user with a doctor or hospital name. Chatbots are machines and are trained on some specific data of a specific domain and hence they are not able to understand the symptoms or problems of the user that do not pertain to the same domain.

In the current situation of COVID-19, chatbots are very helpful to tackle two major problems. First, the users can have a remote COVID-19 test on chatbot that gives the likelihood of contracting COVID-19, and second, the user can chat with the chatbot to ask queries related to COVID-19 and mental health issues. The chatbots are very accurate in solving user queries related to specific topics when they are trained with the ML algorithm. Also, psychiatric chatbots are very helpful to users experiencing mental health–related issues. It provides recommendations and actions on specific problems.

This psychiatric chatbot is available in English language. The authors tried to overcome this language constraint and make the chatbot available in Indian languages too. This chatbot has been implemented by using the Seq2Seq transformer algorithm. The contribution of psychiatric chatbot for COVID-19 in the domain of healthcare is significant.

8.7.2 Conclusion

The COVID-19 pandemic has led to widespread economic and social disruption. One in five people is experiencing mental disorders such as anxiety, depression and insomnia. In addition, misinformation about the COVID-19 pandemic is out of control and leads to confusion and false beliefs. The proposed chatbot will provide on-demand, free-of-cost information to mental health patients, which will surely be effective in relieving some of the pain.

Currently, there is an AI- and ML-powered revolution disrupting the health-care industry. In the coming years, we will be seeing massive changes in the way patients are treated. Besides medical research, technological involvement in treatment has been the main factor in making healthcare effective and available to all. We believe that by developing proposed chatbot, we are contributing to this revolution.

8.7.3 Future Work

The future scope of this project is to use proposed chatbot to create a dataset that can be reviewed by medical professionals. The chatbot will be able to constantly monitor the user's depression levels through regular PHQ-9 tests and also keeps tabs on topics and problems in the user's life. This can be converted into a very useful dataset that the user's psychologist can refer to during a face-to-face meeting.

Due to the increasingly remote way in which people are starting to lead their lives, it is safe to say that mental health is a problem that is bound to increase. Lack of companionship, a feeling of discontent and anxiety are very common problems with the current millennial generation. Millennials are also more likely to confide their trust in an e-counselor. The authors believe that proposed chatbots have a massive scope and would play an effective part in dealing with young people's mental health problems.

India is a multicultural and multilinguistic country. Including English, there are 23 Indian official languages. Also, the proposed chatbot is now available in the English language only. Hence, there is a scope to develop proposed chatbot in as many Indian languages as possible. The proposed chatbot will realise its full potential only when it can be used by a majority of Indians.

REFERENCES

Ayanouz, S., Abdelhakim, B. A., & Benhmed, M. (2020). A smart chatbot architecture based NLP and machine learning for health care assistance. *Proceedings of the 3rd International Conference on Networking, Information Systems & Security*, 1. https://doi.org/10.1145/3386723.3387897

Bhirud, N., Tatale, S., Randive, S., & Nahar, S. (2019). A Literature Review on Chatbots in Healthcare Domain. *International Journal of Scientific & Technology Research, 8*(7), 225–231.

Fitzpatrick, K. K., Darcy, A., & Vierhile, M. (2017). Delivering Cognitive Behavior Therapy to Young Adults With Symptoms of Depression and Anxiety Using a Fully Automated Conversational Agent (Woebot): A Randomized Controlled Trial. *JMIR Mental Health, 4*(2), e19. https://doi.org/10.2196/mental.7785

Javed, M., Nagabhushan, P., & Chaudhuri, B. (2015). A direct approach for word and character segmentation in run-length compressed documents with an application to word spotting. *2015 13th International Conference on Document Analysis and Recognition (ICDAR)*, 216–220. https://doi.org/10.1109/icdar.2015.7333755

Jwala, K., Sirisha, G., & Raju, G. P. (2019). Developing a Chatbot using Machine Learning. *International Journal of Recent Technology and Engineering (IJRTE), 8*(1S3), 89–92.

Kataria, P., Rode, K., Jain, A., Dwivedi, P., & Bhingarkar, S. (2018). User Adaptive Chatbot for Mitigating Depression. *International Journal of Pure and Applied Mathematics, 118*(16), 349–361.

Lee, D., Oh, K.-J., & Choi, H.-J. (2017). The chatbot feels you-a counseling service using emotional response generation. *2017 IEEE International Conference on Big Data and Smart Computing (BigComp)*, 437–440, doi: 10.1109/BIGCOMP.2017.7881752

Ling, W., Dyer, C., Black, A. W., & Trancoso, I. (2015). Two/too simple adaptations of Word2Vec for syntax problems. *Proceedings of the 2015 Conference of the North American Chapter of the Association for Computational Linguistics: Human Language Technologies*, 1299–1304. https://doi.org/10.3115/v1/n15-1142

Mathew, R. B., Varghese, S., Joy, S. E., & Alex, S. S. (2019). Chatbot for disease prediction and treatment recommendation using machine learning. *2019 3rd International Conference on Trends in Electronics and Informatics (ICOEI)*, 851–856. https://doi.org/10.1109/icoei.2019.8862707

Menni, C., Valdes, A. M., Freidin, M. B., Sudre, C. H., Nguyen, L. H., Drew, D. A., Ganesh, S., Varsavsky, T., Cardoso, M. J., El-Sayed Moustafa, J. S., Visconti, A., Hysi, P., Bowyer, R. C. E., Mangino, M., Falchi, M., Wolf, J., Ourselin, S., Chan, A. T., Steves, C. J., & Spector, T. D. (2020). Real-Time Tracking of Self-Reported Symptoms to Predict Potential COVID-19. *Nature Medicine, 26*(7), 1037–1040. https://doi.org/10.1038/s41591-020-0916-2

Rahman, A. M., Mamun, A. A., & Islam, A. (2017). Programming challenges of chatbot: Current and future prospective. *2017 IEEE Region 10 Humanitarian Technology Conference (R10-HTC)*, 75–78. https://doi.org/10.1109/r10-htc.2017.8288910

Richardson, L. P., McCauley, E., Grossman, D. C., McCarty, C. A., Richards, J., Russo, J. E., Rockhill, C., & Katon, W. (2010). Evaluation of the Patient Health Questionnaire-9 Item for Detecting Major Depression Among Adolescents. *PEDIATRICS, 126*(6), 1117–1123. https://doi.org/10.1542/peds.2010-0852

Sancheti, R., Upare, S., Bhirud, N. and Tatale, S. (2020). Adaptive Machine Learning Chatbot for Code-Mix Language (English and Hindi). *International Journal of Recent Technology & Engineering, 8(5)*, 3566-3572. https://doi.org/10.35940/ijrte.E6489.018520

Serafini, G., Parmigiani, B., Amerio, A., Aguglia, A., Sher, L., & Amore, M. (2020). The Psychological Impact of COVID-19 on the Mental Health in the General Population. *QJM: An International Journal of Medicine, 113*(8), 531–537. https://doi.org/10.1093/qjmed/hcaa201

Sisodia, D. & Sisodia, D. S. (2018). Prediction of Diabetes Using Classification Algorithms. *Procedia Computer Science, 132*, 1578–1585. https://doi.org/10.1016/j.procs.2018.05.122

Sojasingarayar, A. Seq2Seq AI Chatbot with Attention Mechanism. *arXiv preprint arXiv:2006.02767*, 2020.

Su, M. H., Wu, C. H., Huang, K. Y., Hong, Q. B., & Wang, H. M. (2017). A chatbot using LSTM-based multi-layer embedding for elderly care. *2017 International Conference on Orange Technologies (ICOT)*, 1117. https://doi.org/10.1109/icot.2017.8336091

Sutskever, I., Vinyals, O., & Le, Q. V. (2014). Sequence to Sequence Learning with Neural Networks. *Advances in Neural Information Processing Systems*, *27*, 3104–3112.

Thirugnanam, M. (2013). A Heart Disease Prediction Model using SVM-Decision Trees-Logistic Regression (SDL). *International Journal of Computer Applications in Technology*, *68*, 11–15. https://doi.org/10.5120/11662-7250.

Thomas, J. S. & Thomas, S. (2018). Chatbot Using Gated End-to-End Memory Networks. *International Research Journal of Engineering and Technology*, *5*(03), 3730–3735.

Varshney, D., Ekbal, A., Nagaraja, G. P., Tiwari, M., Gopinath, A. A. M., & Bhattacharyya, P. (2020). Natural language generation using transformer network in an open-domain setting. *International Conference on Applications of Natural Language to Information Systems*, Springer, 82–93.

Vinyals, O. & Le, Q. A Neural Conversational Model. *arXiv preprint arXiv:1506.05869*, 2015.

Yin, J., Chen, Z., Zhou, K., & Yu, C. A deep learning based chatbot for campus psychological therapy. *arXiv preprint arXiv:1910.06707*, 2019.

Yusof, M. K. & Man, M. (2017). Efficiency of JSON for Data Retrieval in Big Data. *Indonesian Journal of Electrical Engineering and Computer Science*, *7*(1), 250. https://doi.org/10.11591/ijeecs.v7.i1.pp250–262

ANNEXURE

SPYDER IDE 4.2.1

Spyder is an integrated development environment (IDE) that executes programs that are written in the Python programming language. Spyder IDE is an open-source platform that can be very helpful in developing intense ML algorithms. It also gives a very smooth experience to the user while running very large libraries. Spyder IDE is an editor with code completion, syntax highlighting and introspection features. It also supports multiple IPython consoles. It also has various plug-ins available that help us in the rapid development of the modules.

JUPYTER NOTEBOOK

Jupyter Notebook is an open-source project created by the nonprofit organisation Project Jupyter. It is a web browser-based IDE for creating various documents and Python scripts. Jupyter Notebook has its own kernels of many languages, but it is mainly used to develop Python projects. We can download the code from Jupyter Notebook in various output formats such as HTML, presentation slides, LaTeX, PDF, Markdowns and Python program.

ANDROID STUDIO 4.1.1

Android Studio is an IDE used to develop for Google's Android operating system. It is based on JetBrains' IntelliJ IDE for Java. Android studio supports android app development in Java and Kotlin. An Android Studio has various features such as

Gradle-based build support android-specific retrofitting and suggests quick fixes in the run-time environment, lint tools to catch performance, usability and version compatibility. It also has various plug-ins that are helping with the faster development of the projects.

SQLite

SQLite is an open-source relational database. SQLite is used to store, manipulate and retrieve the data at any point of time. It is embedded in android and hence no need to perform any extra installations or database setup.

Python Libraries Required

NLTK

Natural Language Toolkit is a Python library specially made for text analysis. This library provides various functions related to text analysis. We can perform text cleaning, tokenisation, text classification, stemming, lemmatisation and tagging using this NLTK. This library is the heart of natural language processing.

TensorFlow

TensorFlow is an open-source Python library specifically made for building machine learning models. This library provides all the functionalities that are required to build the ML model. The TensorFlow library also provides the Keras API, which is specifically used to build transformers.

Pandas

Pandas is an open-source Python library used to handle datasets. We can perform all the operations on the dataset that acts as a data frame in Pandas.

Matplotlib

Matplotlib is an open-source Python library used to plot all types of graphs. This library is very helpful and used in various domains. Using this graph functionality of this library, we can make a decision. Also, we can plot one thing vs another thing to understand the relation between them. This library is very helpful to see the learning rate of the ML algorithm.

JSON

JSON is a format we can use to store, exchange and manipulate data. JSON is written in JavaScript Object Notation, and it contains normal text that has various intents and intent associated text.

Re

To handle regular expressions, Python provides an Re library. Using this library, we can generate a special sequence for our usage. This library helps us to find a set of strings and various patterns.

Flask

Flask is a Python library using which we can run our ML model on the Flask server. A Flask server hosts our an ML model and hence using APIs, we can communicate to the ML model.

Request

This module provides standards for making HTTP requests in Python. It simplifies the complexities of making requests using an API so that developers can focus on interacting with the services.

ANDROID STUDIO LIBRARIES REQUIRED

Retrofit

Retrofit is a class in android that is used to handle the API requests. It makes it easy to retrieve and upload structured data via REST-based web service.

GSON

GSON is a Java library. Using the GSON library, we can convert Java Objects into their JSON representation.

9 An Analysis of Drug-Drug Interactions (DDIs) Using Machine Learning Techniques in the Drug Development Process

Pramoda Patro, Krishna Kumar,
Debabrata Swain and Hanumantha Rao Sama

CONTENTS

9.1   INTRODUCTION

Combining multiple therapies, which can increase drug efficacy by reducing drug toxicities or drug resistance, is a promising approach for treating a multitude of diseases, including hypertension, cancer, asthma and AIDS [1]. Predicting drug-drug interactions (DDIs) is critical to drug development in healthcare. From an economical point of view, experiments and accurate computations are needed in the area of DDIs [2]. DDIs result in unwanted bodily effects, when multiple drugs are administered to patients. DDIs can result in unexpected and unknown pharmacological effects like ADEs (adverse drug events) [3]. Further, DDIs are also responsible for worldwide increased mortality rates. Patients undergo adverse clinical effects, especially when one drug reduces or deactivates the therapeutic effects of other drugs. DDIs may also

DOI: 10.1201/9781003122401-9

increase the toxicity of a drug, thus compromising treatment outcomes of patients. This increasing intensity of DDI reactions has resulted in the construction of a DDI database by automatically extracting DDI information using natural language processing from clinical and medical literature.

Several studies using different techniques have examined similarities in targeted drugs [4]. These studies have predicted possibilities of interactions among drug pairs but failed to describe pharmacological effects on patients [5]. Moreover, it is imperative to have a large volume of details on drugs to forecast side effects by computational methods and such information is unavailable for DDI predictions [6]. Thus, the prediction of DDIs has been applied on approved or investigational drugs, and these methods have not been examined to analyzes drug-food constituent interactions (DFIs), another category of DDIs [7].

Following this introductory section, the next section is a study of related research in MLTs (machine learning techniques) and DLTs (deep learning techniques) in the area of DDIs. Section 9.3 details the proposed method of this research, followed by its experimental results in Section 9.4. This work concludes in Section 9.5.

9.2   LITERATURE REVIEW

MLTs have been widely used in DDI predictions and a few selected studies have been detailed in this section.

The study by Ezzat et al. [8] used dual-matrix factorisation techniques for DDI detections. Their scheme was learnt using graph regularisations. Their scheme assumed non-occurring edge existence while learning, i.e., missing cases, and formulated a pre-processing step to improve detections of new drugs and new targets. This was done by adding edges during intermediate interactions (likelihood scores). Their experimental cross validations achieved satisfactory and enhanced results compared with other three benchmarked methods. Their simulations of new drugs/new targets predicted otherwise unattended interactions using GRMF (graph-regularised matrix factorisation).

DDIs were also detected by combining heterogeneous information by Pathak et al. [9] in their study. The study joined multiple drug information for generating new drug interactions and targets. Their scheme, KronRLS-Stacking, used a stacking-based approach that combined linear models nonlinearly and thus could predict DDIs. Although their proposed algorithm was based on RLS and KronRLS algorithms, heterogeneous drug information was combined and learnt using stacking, an ensemble technique. Their scheme's benchmarks with 17 other methods showed very satisfactory results. Further, their scheme could operate in parallel and distributed environments.

Monteiro et al. [10] used DLTs to obtain one-dimensional protein sequences. Their scheme used CNNs (convolution neural networks) that extracted amino acid sequences, and compounds were extracted using SMILES (simplified molecular-input line-entry system) strings. These two kinds of information were features that expressed local dependencies or patterns which could then be used by FCNNs (fully connected neural networks) for classifications. In experimentations, the proposed scheme's CNNs obtain important data representations in place of traditional descriptors, thus enhancing performances. The study's end-to-end DLT outperformed traditional MLTs by accurate classifications of negative and positive interactions. In Ref. [11], regarding MLTs, the fuzzy rules' parameters are fine-tuned using hybrid ant colony particle swarm optimisation to minimise complexity and search space. In the second layer, similarity-based directional components of data partitioning and cloud data formation are used to create an efficient model based on fuzzy rules extraction. In this method, wavelet functions are used to calculate the weight and bias values of neurons [12]. A suitable hybrid optimisation problem is suggested using parallel and inseparably fuzzy rules. A novel shapeable membership feature with adaptive contour is used to evaluate fuzzy sets [13].

Accurate predictions of DDIs were attempted by Chen et al. [4] in their study in which graph representations were used for learning data. Their novel proposal overcame two issues of data interpretability and accuracy. The study experimented their scheme on small and large DDI dataset in which results showed a higher performance of the model compared with other related approaches in terms of higher precision and recall values. Further, the proposed model was scalable and identified important local atoms with its attention mechanism, thus gaining domain knowledge with interpretability. The study's scheme was also insensitive to pairwise similarities of data found in test datasets and retrieved drug-pair interactions in spite of the considered drug's pairwise similarity that was extremely low.

Text mining was exploited by Yan et al. [14] in their study that used augmented text-mining features for building their LRs (logistic regressions). The scheme enhanced predictions in its discriminations and calibrations. Their synthesised features outperformed other models that trained on structured features by scoring area under the curve (AUC) (96% vs. 91%), sensitivity (90% vs. 82%) and specificity (88% vs. 81%). The study also discussed implications with its intermediate "hidden topics" results while mining texts.

9.3   PROPOSED METHODOLOGY

The proposed improved classification model (ICM) takes the drug's structural information in pairs as inputs for predicting relevant DDI types. This work preprocesses data by eliminating missing features using min-max method, thus producing better features that are subsequently chosen using ABC (artificial bee colony) algorithm. The selected features are then used for predicting DDIs using ICM. This work also uses DBNs (deep belief networks) for classifying data. Further, IBA (improved bat algorithm) of this work is used in optimisations. The overall flow of this proposed work is depicted in Figure 9.1.

9.3.1   REPRESENTATION OF DATA

This research work used both chemical and biological drug descriptions, including their chemical substructures. Information on the reported 192,303 DDIs were described in the form of sentences from Drug Bank. A total of 192,303 DDIs were classified into 99 general sentence structures, each of which describes a specific DDI type. Among the 99 general sentence structures, the 86 general sentence structures (i.e., 86 DDI types) were considered for the development of ICM because they are associated with at least five drug pairs that could be split into training (60% of the dataset; three drug pairs), validation (20%; one drug pair) and test (20%; one drug pair) datasets to develop the ICM model resulting in 192,284 DDIs from 191,878 drug pairs split across 86 DDI types as the DDI dataset.

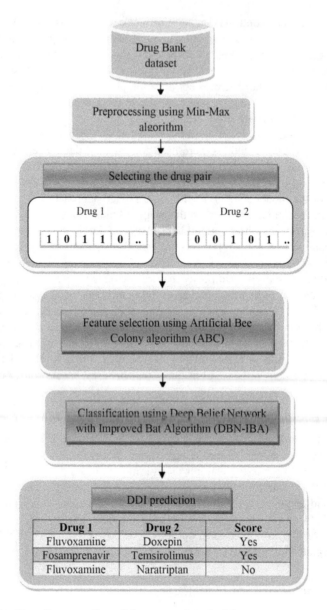

FIGURE 9.1   Overall process flow of the proposed methodology.

9.3.2   PREPROCESSING OF DATA

Data preprocessing is used to assemble useful information groups in a well-refined format from input data. MLTs can be used successfully in predictions or classifications when the quality of input data is high and data preprocessing procedures help in achieving this objective. This work uses min-max data normalisation as its preprocessing procedure to improve the accuracy of DDI predictions.

9.3.2.1 Min-Max Normalisation

This technique normalises input data by transforming them into fixed value ranges using linear transformations [15]. One important quality of this technique lies in its ability to preserve associations between input and scaled values. Generally, in such transformations or scaling, out-of-bound errors occur as normalised values move away from original data values. This work avoids this error by constraining scaled values within a specific range. This normalisation can be depicted in the following equation:

$$X_n = \frac{X_0 - X_{min}}{X_{max} - X_{min}} \tag{9.1}$$

where X_n is the variable X's new value, X_0 is the variable X's current value, X_{min} is the dataset's minimum value and X_{max} is the dataset's maximum value.

9.3.3 ABC-BASED FEATURE SELECTION

Feature selection is the process of determining highly correlated features within the feature set. ABC algorithm's first half encompasses employed bees while the remaining are onlooker bees [16], thus making them equal in proportions within a swarm.

The initial population of ABC is generated by random distribution for SN (swarm size) solutions (food sources). Assuming $X_i = \{x_{i,1}, x_{i,2}, ..., x_i, D\}$ is the ith solution in the swarm, then D is the dimension. Employed bees in X_i generate candidate solutions V_i within the neighborhood and are presented in the following equation:

$$v_{i,j} = x_{i,j} + \Phi_{i,j} \cdot (x_{i,j} - x_{k,j}) \tag{9.2}$$

where $v_{i,j}$ is the selected random candidate solution $(i = k)$, j the random dimension index from $\{1, 2, ..., D\}$ and $\phi_{i,j}$ the random number in the interval $[-1, 1]$. A greedy selection is used for the generation of the new solution V_i. When V_i's fitness value is better than the parent X_i, V_i's value is updated for X_i, else X_i value is unchanged. On the completion of the search process by all employee bees, food source information is shared with onlooker bees using waggle dances. Onlooker bees evaluate the received food information of employee bees and select the food source related to its nectar value. The onlooker bee's probabilistic selection is based on roulette wheel concept and is depicted as

$$P_i = \frac{fit_i}{\sum_{j=1}^{SN} fit_j} \tag{9.3}$$

where fit_i is the ith solution's fitness value in the swarm and solution i has a higher probability of being selected as the ith food source. When positional values do not improve even after a predefined number of iterations, the food source is abandoned.

Assuming X_i is abandoned, scout bees discover new food sources in place of X_i and are defined as

$$x_{i,j} = lb_j + \text{rand}(0,1) \cdot (ub_j - lb_j) \qquad (9.4)$$

where rand(0,1) is the normally distributed random number in the interval [0, 1], lb the lower boundary of jth dimension and ub the upper boundary of jth dimension.

9.3.4 PROPOSED ICM USING DBNs

DBNs encompass multiple layers for classifications. In a feature set F, *fed into DBNs* [17], each layer is made of neurons. DBNs have inputs, hidden and output layers where hidden and input neurons are connected. The connection between visible and hidden neurons is exclusively symmetric where neuron models determine the accuracy of outputs from inputs. Since the output neurons are stochastic and probabilistic in the Boltzmann network, Equation (9.8) depicts the output while Equation (9.9) depicts the sigmoid-shaped function's probability and t^p denotes pseudo-temperature. This stochastic deterministic approach is depicted as in Equation (9.10)

$$O_q(\zeta) = \frac{1}{1 + e^{\frac{-\zeta}{t^p}}} \qquad (9.5)$$

$$\overline{PR} = \begin{cases} 1 \text{ with } 1 - \overline{O}_q(\zeta) \\ 0 \text{ with } \overline{O}_q(\zeta) \end{cases} \qquad (9.6)$$

$$\lim_{t^p \to 0^+} \overline{O}_q(\zeta) = \lim_{t^p \to 0^+} \frac{1}{1 + e^{\frac{-\zeta}{t^p}}} = \begin{cases} 0 \text{ for } \zeta < 0 \\ \frac{1}{2} \text{ for } \zeta = 0 \\ 1 \text{ for } \zeta > 0 \end{cases} \qquad (9.7)$$

Feature extractions are executed using RBM (restricted Boltzmann machine) layers while MLPs (multi-layer perceptrons) classify information. RBM's mathematical approaches use Boltzmann machine's energy for neuron formations or binary state bi, as given in the following equation, where $w_{a,l}$ denotes weights between neurons and θ_a the biases:

$$\Delta E(bi_a) = \sum_l bi_a w_{a,l} + \theta_a \qquad (9.8)$$

The computable energy of visible and hidden neurons (x,y) is depicted in Equations (9.9)–(9.11). x_a represents visible unit's neuron state, W_l represents binary state of l hidden units and K_a, w_a represent network's biases.

$$E(x,y) = \sum_{(a,l)} w_{a,l} X_a Y_l - \sum_a K_a X_a - \sum_l W_l Y_a \qquad (9.9)$$

$$\Delta E(x_a, \bar{y}) = \sum_l w_{al} Y_l + K_a \qquad (9.10)$$

$$\Delta E(\bar{x}, y_a) = \sum_l w_{al} x_a + W_l \qquad (9.11)$$

The probable distribution of inputs is encoded as weights (parameters) and assigned by RBM while learning. RBM assignment of weights to inputs in learning can be depicted as

$$\hat{W}_{(\hat{M})} = \max_{\hat{W}} \prod_{\bar{x} \in; N} c(\bar{x}) \qquad (9.12)$$

For the visible and hidden pair vectors $c(\bar{x}, \overrightarrow{hi})$, RBM's assignment probability can be defined as in Equation (9.13), where PA^F *implies* partitioning function and depicted as Equation (9.14):

$$c(\bar{x}, \overrightarrow{hi}) = \frac{1}{PA^F} e^{-E(\bar{x}, \bar{y})} \qquad (9.13)$$

$$PA^F = \sum_{\bar{x}, \bar{y}} e^{-E(\bar{x}, \bar{y})} \qquad (9.14)$$

This proposed work uses CDs (contrastive divergences)-based learning for achieving sampling expectations in the complex distribution process. However, finding optimal parameters of DBN is a difficult task. This is because the search process is trapped in local minima. Therefore, the accuracy of DBN classification is degraded due to the above-mentioned problem of inaccurate parameters of neural network. So this work introduces an IBA for solving this problem.

9.3.4.1 Improved Identical Distribution-Based Bat Algorithm (IIDBA)

BA is an optimisation algorithm for swarm intelligence and mimics bat's foraging behavior. BA uses the echolocation quality of bats [18]. In the IBA, dataset is considered the initial Bat population and each data has a certain frequency f_i and velocity v_i. The frequency and velocity are estimated using the following equations and updated after each iteration:

$$f_i = f_{min} + (f_{max} - f_{min})\delta \qquad (9.15)$$

$$v_i^t = v_i^{t-1} + \left(G_i^t - G_{\text{current}}\right)f_i \tag{9.16}$$

$$\delta = \delta_0 \exp(-\mu k) \tag{9.17}$$

where δ_0 and μ are constants and k is the current generation, and G_{current} represents the present global solution and δ denotes the uniform distribution function that ranges between 0 and 1.

Then, local search is performed to determine the best solutions using random walks within the dataset. Random best solutions are identified by a simple random sampling technique to extract the predominant features from the entire dataset that contains a large number of features. The features are grouped together based on the similarity and ranked by the resampling method. The current solution is compared with the ranked solutions and they are sorted in the best order. The solutions are updated using the following equation:

$$G_{\text{best}} = G_{\text{current}} + \in S_i \tag{9.18}$$

where $\in$ is the random number in the interval $[-1]$ signified by S_i.

9.4   RESULTS AND DISCUSSION

Datasets were collected for evaluating the proposed scheme's DDI predictions. The Drug Bank dataset has 191,878 drug pairs with DDI details, but only 0.2% (406 drug pairs) belongs to two DDI types, while a major portion 99.8% of the dataset belongs to a single DDI type.

The proposed scheme was evaluated further with Deep DDI outputs (182 drug pairs of Drugs.com dataset found at https://www.drugs.com/) for the consistency of DDI descriptions [19]. This additional DDI information was used with Drug Bank (Dataset S3). DDIs of 182 drug pairs were most frequent (reduced metabolism due to drug interactions where 69 of DDI type 6), serum concentration (37 of DDI type 10), increasing corrected QT interval prolonging activity (23 DDI type 76) and increasing anticoagulant activity of drug interactions (13 DDI type 46).

Experimental results also suggested increased bioavailability (DDI types 6 and 10) as one of the major causes for ADEs in drug pairs (21).

9.4.1   EVALUATION METRICS

Several criteria used for evaluating binary classifications were used in this work for DDI prediction evaluations. The proposed system was evaluated using precision (proportion of relevant retrieved instances) and recall (proportion of relevant instances retrieved accurately).

Though these evaluation metrics conflict with each other, they are combined with equal weights for obtaining an F-measure evaluation metric. This work was also evaluated using accuracy defined as accurately predicted instances relative to total predicted instances.

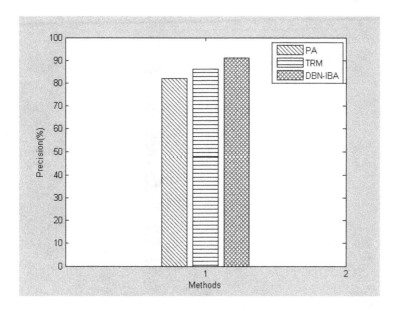

FIGURE 9.2 Precision comparison results between DBN-IBA classification and existing techniques for DDI prediction.

Figure 9.2 illustrates precision comparison results between this work's DBN-IBA classification and existing techniques in DDI predictions. From the results, it concludes that the proposed DBN-IBA technique has high-precision results compared with the existing predicting techniques.

Figure 9.3 illustrates recall comparison results between this work's DBN-IBA classification and existing techniques in DDI predictions. From the results, it concludes that the proposed DBN-IBA technique has high recall results compared with the existing predicting techniques.

Figure 9.4 illustrates *F*-measure comparison results between this work's DBN-IBA classification and existing techniques in DDI predictions. From the results, it concludes that the proposed DBN-IBA technique has high *F*-measure results compared with the existing predicting techniques.

Figure 9.5 illustrates the accuracy comparison results between the proposed and existing techniques for the prediction of DDI using the proposed DBN-IBA classification. From the results, it concludes that the proposed DBN-IBA technique has high-accuracy results compared with the existing predicting techniques.

9.5 CONCLUSION

This work proposed a computational framework that predicts DDIs through a structure similarity profile-based ICM. This work preprocesses data by eliminating missing features using min-max method, thus producing better features that are subsequently chosen using ABC algorithm. The selected features are then used for predicting DDIs using ICM. This work also uses DBNs for classifying data. Further,

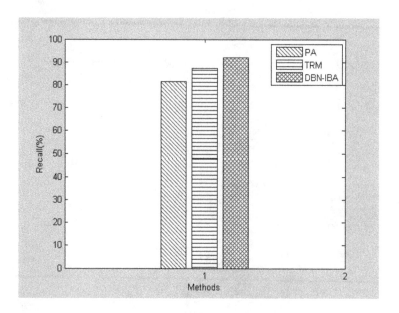

FIGURE 9.3 Recall comparison results between DBN-IBA classification and existing techniques for DDI prediction.

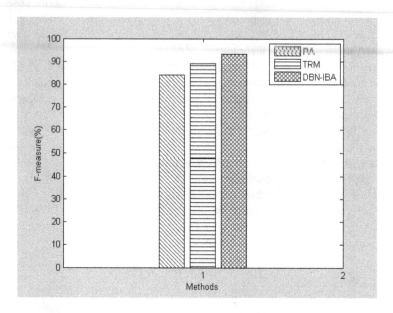

FIGURE 9.4 *F*-measure comparison results between DBN-IBA classification and existing techniques for DDI prediction.

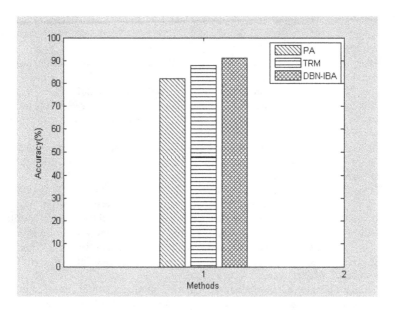

FIGURE 9.5   Accuracy comparison results between DBN-IBA classification and existing techniques for DDI prediction.

IBA of this work is used in optimisations. The proposed DBN-IBA model's simulations on MATLAB and subsequent evaluations with performance metrics show the model's effectiveness by displaying better DDI prediction rates. The collected database and proposed method can be used as a powerful pharmacovigilance tool and also in combination with other methods like hybrid learning and efficient feature extraction techniques.

REFERENCES

1. Xu, B., Shi, X., Zhao, Z., & Zheng, W. Leveraging biomedical resources in Bi-LSTM for drug-drug interaction extraction. *IEEE Access, 6*, 33432–33439, 2018.
2. Dewi, I. N., Dong, S., & Hu, J. Drug-drug interaction relation extraction with deep convolutional neural networks. In *2017 IEEE international conference on bioinformatics and biomedicine (BIBM)* (pp. 1795–1802). IEEE, 2017.
3. Ren, Y., Fei, H., & Ji, D. Drug-drug interaction extraction using a span-based neural network model. In *2019 IEEE international conference on bioinformatics and biomedicine (BIBM)* (pp. 1237–1239). IEEE, 2019.
4. Chen, X., Liu, X., & Wu, J. Drug-drug interaction prediction with graph representation learning. In *2019 IEEE international conference on bioinformatics and biomedicine (BIBM)* (pp. 354–361). IEEE, 2019.
5. Hu, P. W., Chan, K. C., & You, Z. H. Large-scale prediction of drug-target interactions from deep representations. In *2016 International joint conference on neural networks (IJCNN)* (pp. 1236–1243). IEEE, 2016.
6. Mongia, A., & Majumdar, A. Deep matrix completion on graphs: Application in drug target interaction prediction. In *ICASSP 2020–2020 IEEE international conference on acoustics, speech and signal processing (ICASSP)* (pp. 1324–1328). IEEE, 2020.

7. Fakhraei, S., Huang, B., Raschid, L., & Getoor, L. Network-based drug-target interaction prediction with probabilistic soft logic. *IEEE/ACM Transactions on Computational Biology and Bioinformatics, 11*(5), 775–787, 2014.
8. Ezzat, A., Zhao, P., Wu, M., Li, X. L., & Kwoh, C. K. Drug-target interaction prediction with graph regularized matrix factorization. *IEEE/ACM Transactions on Computational Biology and Bioinformatics, 14*(3), 646–656, 2016.
9. Pathak, S., & Cai, X. Ensemble learning algorithm for drug-target interaction prediction. In *2017 IEEE 7th international conference on computational advances in Bio and medical sciences (ICCABS)* (pp. 1–1). IEEE, 2017.
10. Monteiro NRC, Ribeiro B, Arrais J. Drug-Target Interaction Prediction: End-to-End Deep Learning Approach. *IEEE/ACM Trans Comput Biol Bioinform.* 2020 Feb 28. doi: 10.1109/TCBB.2020.2977335. Epub ahead of print. PMID: 32142454.
11. Patro, P., Kumar, K., & Suresh Kumar, G. Neuro fuzzy system with hybrid ant colony particle swarm optimization (HASO) and robust activation. *Journal of Advanced Research in Dynamical and Control Systems, 12*(03-Special Issue), 741–750, 2020.
12. Patro, P., Kumar, K., Suresh Kumar, G., et al. Similarity and wavelet transform based data partitioning and parameter learning for fuzzy neural network. *Journal of King Saud University – Computer and Information Sciences,* 2020. https://doi.org/10.1016/j.jksuci.2020.06.003.
13. Patro, P., Kumar, K., & Suresh Kumar, G. Optimized hybridization of ant colony optimization and genetic algorithm (HACOGA) based IC-FNN classifier for abalone. *Journal of Computational and Theoretical Nanoscience, 17*, 2756–2764, 2020.
14. Yan, S., Jiang, X., & Chen, Y. Text mining driven drug-drug interaction detection. In *2013 IEEE international conference on bioinformatics and biomedicine* (pp. 349–354). IEEE, 2013.
15. Pinto, P. Introducing the min-max algorithm. *Submitted to the AI Depot Article Contest,* 1–10, 2002.
16. Gao, W. F., Huang, L. L., Liu, S. Y., & Dai, C. Artificial bee colony algorithm based on information learning. *IEEE Transactions on Cybernetics, 45*(12), 2827–2839, 2015.
17. Salakhutdinov, R., & Murray, I. On the quantitative analysis of deep belief networks. In *Proceedings of the 25th international conference on machine learning* (pp. 872–879), 2008.
18. Fister, I., Yang, X. S., Fong, S., & Zhuang, Y. Bat algorithm: Recent advances. In *2014 IEEE 15th international symposium on computational intelligence and informatics (CINTI)* (pp. 163–167). IEEE, 2014.
19. Ryu, J. Y., Kim, H. U., & Lee, S. Y. Deep learning improves prediction of drug–drug and drug–food interactions. *Proceedings of the National Academy of Sciences, 115*(18), E4304–E4311, 2018.

10 Image Processing-Based Fire Detection Using IoT Devices

Souvik Das, Jyotirmoy Das, O.B. Krishna and J. Maiti

CONTENTS

10.1 INTRODUCTION

Most warehouses have complex electrical circuits and machinery that could induce a fire in the workspace through malfunctioning. A fire in a warehouse or factory can lead to huge disruption and incur high costs to a company. Hence, a dependable and automated fire detection system is paramount in factory. Conventionally, fire detection methods require dedicated personnel to do round or implement devices. Usually, smoke sensors are used for this purpose. The alarms are set if a proper buildup of smoke is made. In addition, there are chances that it may be unable to detect the smoke because of wind directions. They rely on chemical reactions and hence are susceptible to false-positive alarms. These sensors become highly unreliable in a large warehouse with a variety of activities tacking place. Implementation also has a high cost. Hence an image on a vision-based system helps in real-time detection and can also reduce false alarms. In addition to this, it can also act alone and give a preview of the situation to the responsible authorities.

Therefore, the current study has proposed an IoT-based (where IoT is the Internet of Things) system in which a camera would be connected to a server and would collect data every 3 minutes. Each image will be processed in real time and if an abnormality is detected using the deep residual learning algorithm, a warning message will be sent to the concerned authorities and the fire safety protocol such as sprinklers will be implemented. The image will also help the fire department personnel to assess where exactly the fire is originated and hence can be helpful in saving life and property.

In a camera or vision-based system, image processing is the key principle. This principle of a camera base fire safety IoT implementation has been supplemented by CCTV cameras that already have been implemented in the workspace for security concerns. Hence these cameras can be modified and become an important tool in protecting a warehouse against fire. Since a person alone cannot keep an eye on more than 20 cameras in a large warehouse, a camera equipped with image processing can alert the personnel and hence real-time detection can be possible.

In this study, we have used deep residual learning for image processing purpose. Deeper neural networks become increasing tougher to test and train. We have implemented a residual learning framework to basically obtain better results on much deeper networks than performed earlier. Hence to train a deep neural network, this technique was used and we were able to establish a classification model by training on a dataset.

The current study provides an improved framework to use IoT and machine learning (ML) techniques for credible decision-making and extends the concept of Industry 4.0 toward risk reduction and safety management. The study strengthens the literature on Industry 4.0 and IoT-based safety management principle.

Section 10.2 gives an overview of the literature related to vision-based fire safety in the industry. Section 10.3 provides the framework of the IoT-based system. Section 10.4 explains the used deep learning algorithm. Section 10.5 represents the results and the code. Section 10.6 concludes the report and showcases areas of future research.

10.2　LITERATURE REVIEW

The proposed framework in the study is expected for smoke or fire identification in automobiles. Sensory devices such as LM35 and CO_2 sensory devices are utilised for detecting fire. Hence, this strategy will recognise fire within a few minutes and air-conditioning framework are utilised for extinguishing the fire. Fuzi et al. (2014) proposed a framework for detecting and recognising timberland fire with the assistance of remote sensor arrangement. The author actualised how to take a shot at the information gathered by sensors instead of detecting the shoot. For preparing and making the system vitality effective for gathered information, they utilised an unbiased system. In light of the video-tapes or video data from cameras, a fire-disallowing framework has been made (Kwon et al., 2008). Islam et al. (2015) provided a framework enabling a system to detect fire and checking the fire by utilising different sensor devices for detecting substances such as CO_2, temperature for detecting fire, and triggering a fire safety protocol.

Distinguishing fire is a significant issue, current innovation is in critical need of a fitting identification framework that can diminish the harm caused due to an enormous number of fire mishaps occurring ordinary (Mueller et al., 2013). At first, scientists endeavored to create carefully assembled strategies for fire origin through concentrating on the movement and shading homes of the fire identification. An image processing model is proposed by Chen et al. (2004) where both the spectrum quality of fire and its movement as inputs are used to identify real-life fire identification. In another work, Çelik et al. (2007) attempted to understand fire and smoke in foggy conditions using a couple of diverse shading areas and so as to formulate the characterisation increasingly, they embraced ideas from fuzzy purpose to segregate fire and smoke from similar areas. Whereas Rafiee et al. (2011) utilised an RGB shading house for fireplace location, Qiu et al. (2012) utilised the YCbCr shading area and made a few adjustments to defeat the disadvantage from the previous approach with the aid of making some gradually nonexclusive standards to become aware of the fire, but excessive bogus identification fee and hindrance of place simply at a weak separation have been the related disadvantages. The shading property has advantages; motion has to be built in the paradigm to distinguish the fire in sure works. Rafiee et al. (2011) utilised a steady and dynamic house of the fire and smoke. However, the charges stay a problem right here likewise because of the nearness of specific protests based on the same shading homes as the fire pixelated squares in the image. Qiu et al. (2012) framed an auto-versatile aspect cognizance calculation for fireplace recognition. Rinsurongkawong et al. (2012) utilised the changing environment in homes of the fire for fireplace identification; however, this approach likewise fizzled with photos having an actual fireplace like articles out of sight. This drawback was that they built a framework on optical movement detectors for isolating furnace from the neutral things. Mobin et al. (2016) presented a fire-place area framework that uses one-of-a-kind sensors to recognise fire, and par-ticularly smoke. Be that as it may, the utilisation of various sensors precipitated the framework is to be gradually costly. Zhang et al. (2016) showed in his work the woodland fire consciousness where they used fireplace patches discovery with an adjusted pretrained CNN (Krizhevsky & Hinton, 2012), while Sharma et al. (2017) proposed a neural network-based furnace region method using ML algorithm such as VG16 (Simonyan & Zisserman, 2015) and ResNet60 (He et al., 2015) as sample designs.

10.3   PROBLEM DESCRIPTION

Disadvantages of smoke, temperature, or light sensors:

1. Are slow in response time
2. Do not provide adequate information about the fire
3. Are costly to implement in a huge warehouse
4. Are unreliable and can give rise to false alarms

Hence, we have built an IoT system that uses a camera to detect an image every 2 minutes. It processes the image in real time and provides a warning if it observes any anomaly in the image, for example, a fire or smoke. Accordingly, the image and warning sign will be sent to the responsible authorities.

Advantages:

1. Gives a real-time response
2. Can provide a 2-D preview of the situation
3. Have to just check the image in case of false alarm
4. Can cover a large area and hence is inexpensive
5. Can be integrated into already installed surveillance cameras

10.3.1   SETUP

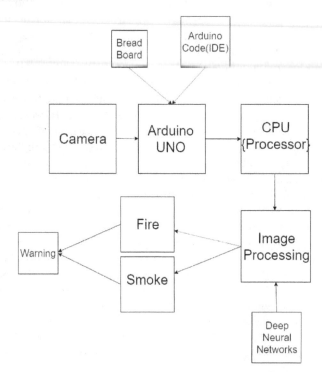

FIGURE 10.1   Image processing fire-based fire detection.

10.3.2 Parts Description

10.3.2.1 Arduino UNO (Main Board)

The Arduino UNO principle board is an electronic small-level computational element dependent on a simple input/output connection and a processing unit domain that utilise the handling. It can be used standalone after a codded program is uploaded or it may be wired to programming on the PC very well. It consequently selected the power supply either from USB or outer power. This board is uncovered the microcontroller input/yield stick to empower association between this stick to other circuit or sensor. An important advantage of the Arduino is the default way that fittings are modeled, giving the CU a chance to board to be wired to all sort of gainful add-on modules called programs. Some of the circuits worked with the board dependent on different connections; however, a great deal of shields is autonomously worked by means of an I²C sequential circuit, letting numerous connections to consolidate and associated in parallel. Most of them are controlled by the UNO AVR system of chips particularly the Arduino UNO. Some processors are made to be Arduino compatible. Heaps of circuits are outfitted with a 5-V direct controller and a 16-MHz oscillator (in certain ways), anyway some creation for model the LilyPad is controlled by 8 MHz.

10.3.2.2 Arduino Software

The Arduino IDE platform is an operation-based application coded in Java and is made from the basic platform designed in processing programming language and circuit diagrams. It is a creation to instruct coding to regular citizen and other new gatherings uncommon with coding betterment. It comprises a code editorial manager with trademark, for instance, punctuation suggestion, recommendation system, and programmable space, and is ready to pack and introduce code projects into the circuit board with an instant push of a button. There is just not required to change the basic coding on a direction line platform. Making on direction line is simple can be improved whenever the need arises using external hardware. The Arduino IDE is based on a C++ library named "Wiring," which take bunches of information with a lot less difficultly. Arduino programs are coded in C/C++, and programmers need to specify two main programs to make it work:

- setup ()—It runs once and is responsible for the predefined definitions and instructions that will be needed in the code.
- loop ()—Its function is to repeat a set of commands again and again against a constraint.

It is an attribute of most Arduino sheets that they have a LED and burden-resistor wired between stick 13 and ground, a helpful component for some simple tests.

10.3.3 Connections

FIGURE 10.2   Aurdino UNO or any other IOT devices.

FIGURE 10.3   OV7670 Camera sensor module was used in this project.

FIGURE 10.4 Resistors 4 pcs (2 pcs 10 KOhm, 2 pcs 4.7 KOhm)

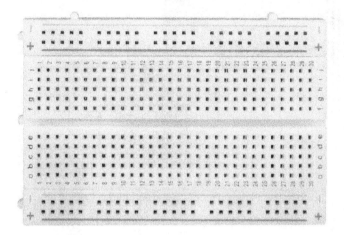

FIGURE 10.5 Bread board half-size or small size 1 pc.

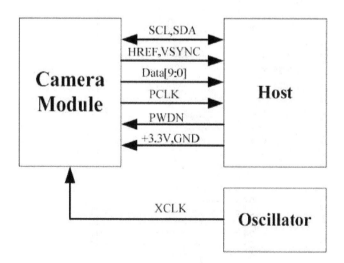

FIGURE 10.6 Camera, host and oscillator.

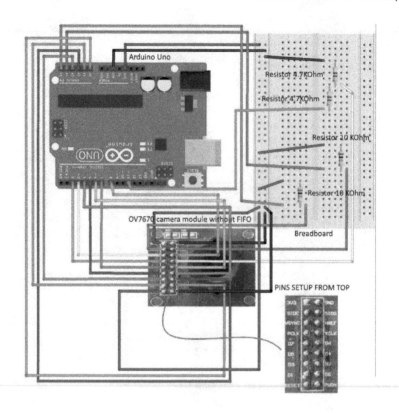

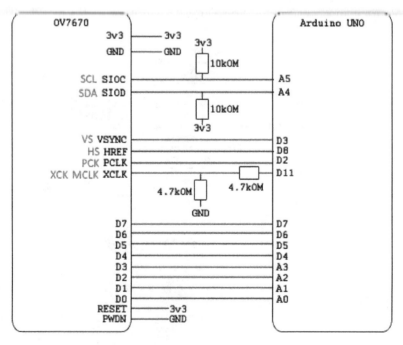

FIGURE 10.7   Connections camera module, Arduino UNO.

FIGURE 10.8   Image-processing fire detection setup.

10.4   METHODOLOGY

10.4.1   Residual Learning (**RSNET** Model)

This model was developed by Kaiming He et al. where they proposed a deep neural network-based model using techniques such as deep residual learning to shorten the connections between the deeper layers to the answers and hence get a much better network through this optimisation.

10.4.2   Code Snippets and Explanation

```
[4]  from PIL import Image

[5]  from torchfusion_utils.fp16 import convertToFP16
     from torchfusion_utils.initializers import *
     from torchfusion_utils.metrics import Accuracy
     from torchfusion_utils.models import load_model,save_model

[6]  import time
     import torch
     import torch.nn as nn
     import torch.nn.functional as F
     import torchvision
     from torchvision import datasets, transforms, models
     import numpy as np
     import matplotlib.pyplot as plt
     import matplotlib.pyplot as plt
     from torch.autograd import Variable
```

10.4.2.1   Code Part 1 Explanation

1. PyTorch and necessary libraries were installed to do image processing.
2. It was done in Jupyter Notebooks on Google Colab.

```
[7]  transforms_train = transforms.Compose([transforms.Resize(225),
                                transforms.CenterCrop(224),
                                transforms.ToTensor(),
                                transforms.Normalize([0.5, 0.5, 0.5],
                                            [0.5, 0.5, 0.5])])

     transforms_test = transforms.Compose([transforms.Resize(225),
                                transforms.CenterCrop(224),
                                transforms.ToTensor(),
                                transforms.Normalize([0.5, 0.5, 0.5],
                                            [0.5, 0.5, 0.5])])

     batch_sizes = 64
     test_data_dir = './FIRE-SMOKE-DATASET/Test'
     train_data_dir = './FIRE-SMOKE-DATASET/Train'

     train_data = datasets.ImageFolder(root=train_data_dir, transform=transforms_train)
     test_data = datasets.ImageFolder(root=test_data_dir, transform=transforms_test)

     train_data_loader = torch.utils.data.DataLoader(train_data, batch_size=batch_sizes, shuffle=True)
     test_data_loader = torch.utils.data.DataLoader(test_data, batch_size=batch_sizes, shuffle=True)
```

10.4.2.2   Code Part 2 Explanation

1. Here, we first download the datasets on to the Google Colab.
2. The batch size is defined as 64 for this.
3. The data is split into two datasets
4. And finally, it's loaded onto the PyTorch.

```
[9]  ResNet = models.resnet50(num_classes=3)

[10] device = torch.device('cuda:0' if torch.cuda.is_available() else 'cpu')
```

10.4.2.3   Code Part 3 Explanation

1. ResNet is the basic algorithm that will train on the dataset and give the results.
2. Here the number of classes is three, "SMOKE," "FIRE," "NEUTRAL."

```
[11] Model = ResNet

     Model = Model.to(device)

     lr = 0.001

     criteria = nn.CrossEntropyLoss()

     optimizer = torch.optim.Adam(Model.parameters(), lr=lr)

     Model,optimizer = convertToFP16(Model, optimizer)

     milestones = [100, 150]

     scheduler = torch.optim.lr_scheduler.MultiStepLR(optimizer, milestones, gamma=0.1)
```

10.4.2.4   Code Part 4 Explanation

1. This shows the model parameters that were used.

```
def model_traing_and_validation_loop(Model, n_epochs, save_path):

    n_epochs = n_epochs

    saving_criteria_of_model = 0

    training_loss_array = []

    validation_loss_array = []

    train_acc = Accuracy()

    validation_acc = Accuracy(topK=1)

    for i in range(n_epochs):

        total_test_data = 0

        total_train_data = 0

        correct_test_data = 0

        training_loss = 0

        validation_loss = 0

        train_acc.reset()
```

```
[2]         for data, target in train_data_loader:

                data, target = data.to(device), target.to(device)

                optimizer.zero_grad()

                predictions = Model(data)

                loss = criteria(predictions, target)

                optimizer.backward(loss)

                optimizer.step()

                training_loss += loss.item()*data.size(0)

                train_acc.update(predictions, target)

            scheduler.step()

        with torch.no_grad():

            validation_acc.reset()

            for data, target in test_data_loader:

                data, target = data.to(device), target.to(device)
```

```
            predictions = Model(data)
                loss = criteria(predictions, target)
                validation_acc.update(predictions, target)
                total_test_data += target.size(0)
                validation_loss += loss.item()*data.size(0)

        training_loss = training_loss / len(train_data)
        validation_loss = validation_loss / total_test_data
        training_loss_array.append(training_loss)
        validation_loss_array.append(validation_loss)

        print(f'{i+1} / {n_epochs} Training loss: {training_loss}, Tran_Accuracy: {train_acc.getValue()}, Validation_loss: {validation_loss}, Validation_Accura
        if saving_criteria_of_model < validation_acc.getValue():
            torch.save(Model, save_path)
            saving_criteria_of_model = validation_acc.getValue()
```

10.4.2.5 Code Part 5 Explanation

This is the main function in which the model is trained.

1. Here the validation loss and train accuracy are reduced on each iteration.
2. Epoch means number of iterations.
3. If model is with a validation loss less, the parameter values are stored on two the computer.

```
def inferenceing_function(img_path):

    def image_display(img_path):

        plt.figure(figsize=(30, 6))
        plt.imshow(Image.open(img_path))
        plt.yticks([])
        plt.xticks([])

        plt.show()
        return 'Image of:'

    def model_inference_results():

        transformer = transforms.Compose([transforms.Resize(225),
                                    transforms.CenterCrop(224),
                                    transforms.ToTensor(),
                                    transforms.Normalize([0.5, 0.5, 0.5],
                                                [0.5, 0.5, 0.5])])
        img = Image.open(img_path)
        img_processed = transformer(img).unsqueeze(0)
        img_var = Variable(img_processed, requires_grad= False)
        img_var = img_var.cuda()
        load_saved_model.eval()
        logp = load_saved_model(img_var)
        expp = torch.softmax(logp, dim=1)
        confidence, clas = expp.topk(1, dim=1)

        return f'Class: {class_name[clas]}', f'Confidence score: {confidence.item()}'
```

10.4.2.6 Code Part 6 Explanation

1. This is basically the prediction function.
2. An image path is given as an input.
3. It traverses through the path and loads the current image.
4. Transforms it then ML model is trained on it.
5. It gives output to the class and the confidence score.

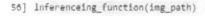

('Image of:', 'Class: SMOKE', 'Confidence score: 0.9755859375')

10.5 RESULTS

Here we will showcase the performance of the algorithm in three different cases.

1. For a neutral picture and accordingly the confidence score:

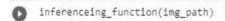

('Image of:', 'Class: Neutral', 'Confidence score: 0.8193359375')

2. A picture with fire:

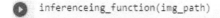

```
inferenceing_function(img_path)
```

```
('Image of:', 'Class: FIRE', 'Confidence score: 1.0')
```

3. A picture with only smoke:

```
img_path = './fire-smoke-neutral-inference-images/smoke/image_15.jpg'
inferenceing_function(img_path)
```

```
('Image of:', 'Class: SMOKE', 'Confidence score: 0.9951171875')
```

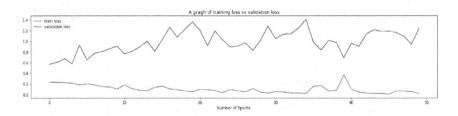

FIGURE 10.9   This figure shows how the training loss reduces

10.6   CONCLUSION

From this report, we can conclude that the image processing can be viably implemented using IoT device and would be instrumental against fire. It also showcases and explains that traditional methods are effective but an image detection provides much more feasibility on the large scale. IoT devices couple with image processing algorithms can play pivotal role in safety of a factory. Hence, they should be implemented.

10.6.1   FUTURE RESEARCH

1. Fire detection through a video output
2. General abnormality detection
3. Implementation in factories that have in general fire and smoke as a background

REFERENCES

Çelik, T., Özkaramanlı, H., & Demirel, H. (2007). Fire and Smoke Detection without Sensors: Image Processing Based Approach. *2007 15th European Signal Processing Conference*, 1794–1798.

Chen, T.-H., Wu, P.-H., & Chiou, Y.-C. (2004). An Early Fire-Detection Method Based on Image Processing. *2004 International Conference on Image Processing, 2004. ICIP'04*, 1707–1710.

Fuzi, M. F. M., Ibrahim, A. F., Ismail, M. H., & Ab Halim, N. S. (2014). HOME FADS: A Dedicated Fire Alert Detection System Using ZigBee Wireless Network. *2014 IEEE 5th Control and System Graduate Research Colloquium*, 53–58.

He, K., Zhang, X., Ren, S., & Sun, J. (2015). Deep Residual Learning for Image Recognition. *CVPR*, 1–9.

Islam, T., Rahman, H. A., & Syrus, M. A. (2015). Fire Detection System with Indoor Localization using ZigBee Based Wireless Sensor Network. *2015 International Conference on Informatics, Electronics\& Vision (ICIEV)*, 1–6.

Krizhevsky, A., & Hinton, G. E. (2012). ImageNet Classification with Deep Convolutional Neural Networks. *Advances in Neural Information Processing Systems*, 1–9.

Kwon, O.-H., Cho, S.-M., & Hwang, S.-M. (2008). Design and Implementation of Fire Detection System. *2008 Advanced Software Engineering and Its Applications*, 233–236.

Mobin, I., Islam, N., & Hasan, R. (2016). An Intelligent Fire Detection and Mitigation System Safe from Fire (SFF). *International Journal of Computer Applications*, *133*(6), 1–7.

Mueller, M., Karasev, P., Kolesov, I., & Tannenbaum, A. (2013). Optical Flow Estimation for Flame Detection in Videos. *IEEE Transactions on Image Processing*, 22(7), 2786–2797.

Qiu, T., Yan, Y., Lu, G., & Member, S. (2012). An Autoadaptive Edge-Detection Algorithm for Flame and Fire Image Processing. *IEEE Transactions on Instrumentation and Measurement*, 61(5), 1486–1493.

Rafiee, A., Dianat, R., Jamshidi, M., Tavakoli, R., & Abbaspour, S. (2011). Fire and Smoke Detection using Wavelet Analysis and Disorder Characteristics. *2011 3rd International Conference on Computer Research and Development*, 262–265.

Rinsurongkawong, S., Ekpanyapong, M., & Dailey, M. N. (2012). Fire Detection for Early Fire Alarm Based on Optical Flow Video Processing. *9th International Conference on Electrical Engineering/Electronics, Computer, Telecommunications and Information Technology, IEEE, 2012*, 1–4.

Sharma, J., Granmo, O.-C., Goodwin, M., & Fidje, J. T. (2017). Deep Convolutional Neural Networks for Fire Detection in Images. *International Conference on Engineering Applications of Neural Network*, Springer, 183–193.

Simonyan, K., & Zisserman, A. (2015). Very Deep Convolutional Networks for Large-Scale Image Recognition, 1–14.

Zhang, Q., Xu, J., Xu, L., & Guo, H. (2016). Deep Convolutional Neural Networks for Forest Fire Detection. *International Forum on Management, Education and Information Technology Application, Ifmeita*, 568–575.

11 Crowd Estimation in Trains by Using Machine Vision

Yogesh Jadhav and Swapnil Pawar

CONTENTS

11.1   INTRODUCTION

11.1.1   MUMBAI SUBURBAN RAILWAYS

Mumbai Suburban Railway (Kusters, 2017, p. 174) is used by the residential and general population of Mumbai. The suburban railway serves almost 7.5 million passengers per day with a total of 2344 trips per week. The Mumbai Suburban Railway is one of the busiest passenger services in the world (2.64 billion). Trains

DOI: 10.1201/9781003122401-11

operate from 4 am to 1:30 am, and some also operate from 2:30 am. It is the first suburban network being operated by Indian Railways.

Mumbai Suburban Railway suffers from the most severe crowding in the country because of its large number of stops and widespread usage by the local urban community.

A 12- or 15-car train has to accommodate around 4,500 people compared to a normal capacity of 2,000, during peak hours. As a result of the cars of 30–50 passengers, a so-called super dense crush charging was sought.

The suburban rail network in Mumbai has been a major reason for deaths of a massive number of people on average ever since it was created. In 2008, there were 17 deaths per weekday due to extensive fatal accidents. The issue of overcrowding is a contributor to injuries and fatalities. People living in Mumbai use these Mumbai Suburban Railway for daily communication. Overcrowding occurs in the trains during peak hours, mostly in the morning, when people travel by these trains to go to their office places and in the evening when they are returning their homes. During these hours, heavy rush prevails in these trains.

A system is needed to identify crowding in particular trains so that people will be able to know about the rush occurring in a particular train and they can decide whether to board the train or not.

11.1.2   What Is Object Detection

Object detection (Viola & Jones, 2001) is a computer vision (Forsyth & Ponce, 2011, pp. 1–3) technique that is used to detect objects. It involves the use of bounding boxes and is used to count objects, track their accurate location and label them. Imagine an image that contains one person and a dog (as shown in Figure 11.1). Object detection enables us to label the objects that are found in the image.

Object detection is distinguishable from image recognition because they are very different activities.

It is used to label images. A picture of dog will be identified as "dog" in the mind. Even though the picture contains two dogs in it, label will still be dog. In object detection,

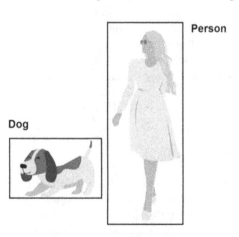

FIGURE 11.1   Object detection.

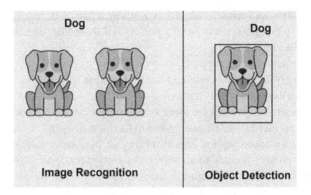

FIGURE 11.2   Image recognition vs. object detection.

it will draw a box around each object. It will also identify where each object is present and label it. It provides more information about the object as opposed to image recognition as shown in Figure 11.2.

11.1.2.1   Types of Object Detection Approaches

Object detection can be accomplished through machine learning or deep learning approaches.

Machine learning uses artificial intelligence (AI) to make an automated decision. Machine learning algorithms rely on historical records to predict future outcomes.

A machine learning approach is based on computer algorithms that help the classification of objects in an image. This feature set is being used to train the model by using regression models. So this will determine object's location in an image with labels.

Another approach is deep learning-based approach. Deep learning is an AI function that mimics the way a human brain operates and works. Deep learning AI allows deep learning without human supervision.

In this, a convolutional neural network (CNN) is applied, which performs end-to-end object detection by using an unsupervised method. In this method, features set no requirements for defining and extracting.

11.1.2.2   Why Object Detection Is Important

Object identification is closely linked to other image recognition and segmentation techniques as it allows us to understand and interpret video or image scenes. There are some variations to the scenes. Machine vision only classifies objects and takes a pixel-level understanding. That's why they are called "identifiers". We can figure out how many objects there are.

Based on these capabilities, object detection is used for

* *Face detection/recognition*
    Face recognition/detection is a computational technique used to recognise human faces inside digital images in a number of applications. Face detection often refers to the behavioral method by which people find and care for faces in a visual environment. Today, many applications are using face recognition algorithms for various purposes such as biometric

identification and whether a mask is worn by a person or not. It is also used to identify facial expressions of a person and, based on that, sentiments of the person are detected.

- *Anomaly detection*

    Anomaly detection is an event-based detection in situation, which works best as an example in this text.

    For example, a model for identifying contaminants in agriculture might identify problems that are not visible to the human eye.

    As entity identification has an ability to help treat patients with specific symptoms, this is also an object recognition model of skin care, which could detect and identify acne in seconds.

    This is something very interesting about these future applications, as they use experts and experience that are normally found in agricultural specialists or physicians.

- *Crowd counting/Estimation*

    An important application of object detection is in the field of crowd counting. This approach will hopefully help in assuring security in transit infrastructures such as malls and amusement parks.

    This tool can help companies to boost infrastructure to position everyone in different locations at any given time. Item identification may also assist communities in organising actions, dedicating public resources, etc. Crowd estimation is very important in terms of calculating person density in a particular area and, according to that, better resource management and planning can be done.

- *Self-driving cars*

    The success of autonomous car prototypes relies on real time detection models. The device must be able to identify, find and trace objects around itself.

    Object detection remains a fundamental task in today's AI for self-driving of cars. But for self-driving, the speed of object detection should be very high and according to that algorithms need to be designed and also the accuracy of object detection is very important in such cases.

- *Video surveillance*

    As object detection techniques allow multiple instances of the given object within a scene to be correctly detected and monitored, these techniques are appropriate for the automation of video surveillance systems.

For example, object recognition models are able to monitor several objects in real time, in a given scene, or through video frames. These kinds of granular surveillance can offer insightful visibility into safety, employee efficiency and safety, retail foot traffic and more from retail to industrial floors.

11.1.3 How Object Detection Works

For object detection, models using deep learning usually require two parts. An encoder uses features of an image to aid search. The predictions made by the encoder to the decoder are transferred to the classifier that confirms them.

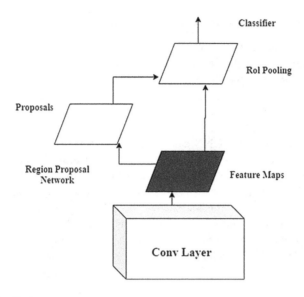

FIGURE 11.3   Basic structure.

A pure return unit is the simplest decoder. The regressor determines the straight distance a regression line assumes. An output of the analysis model is the X, Y coordinate pair and its size. This is a minimalist design. You must specify the amount of bins prior to the delivery. If you have two dogs in your model, only one object can be identified, the other will go unrecognised. However, how many objects you can find will appear in each image by using the regressors. Figure 11.3 shows the basic structure of object detection model.

Then CRN is an application of a regression method. In the decoder model, regions of the image determine what objects exist. Then, the pixels are treated and sent to a bit map (or the idea is rejected). The previous message consists of pixels that are classified into certain areas. This technique can obtain unlimited numbers of region bounding boxes. We are sacrificing computational throughput for the sake of minimising uncertainties.

The Single-Shot detectors are eagerly searching for the light. Whether using a default region, instead of using a subnet Single-Shot Multibox Detection (SSD), it depends on a collection of default regions. A grid is first positioned over the input image, and different boxes of different shapes and sizes are used as regions at each respective anchor point. For each box at each anchor point, the model will determine if there will be an object there; the direction it will move in as well as the size of the box. SSDs are claimed to have the capability to generate many overlapping detections at each node in a distributed fashion. You should consider post-processing SSD outputs to extract the relevant ones. The most common post-processing step is total deletion.

Quite early attempts are made to recognise people from enormous data sets and complex CNN networks. Despite the advances of computer vision algorithms, detection and classification of objects are still not very practical due to the large data requirement. However, the attempt proved futile because it aimed at a high degree of accuracy with repetitive looking rather than just looking once at some points and sensing with a good degree of accuracy.

11.2   RELATED WORK

The main difference between object detection and classification algorithms is that we have to find the bounding box of an object in the detection algorithm and we try to classify the object in the classification algorithm. In situations where you need to draw each object on the screen on one slide, remember that you do not usually need to fill in the box of each object. As you do not know how many objects will be there, do not worry about that.

The main reason why a CNN can't handle the problem is that the output layer (the last layer) is variable. The number of classes of training examples is not necessarily fixed. A naïve approach would be to find the location of the object from the original image and apply a CNN to find the object in the computer. Features offered by some eye-tracking glasses can be different depending on the spatial positions, aspect ratios and image presented. Running this grid in the future will allow a lot of choices based on which types of areas are chosen and where they are located. New techniques have evolved to insure that these high occurrences are recognised faster.

11.2.1   R-CNN, Faster R-CNN and Mask R-CNN

The R-CNN (Girshick et al., 2016, p. 149), that is, a region-based CNN family, contains a variety of common object detection models. These architectures are short for regionally fully CNNs (FNNs) and are based on the above-mentioned region proposal framework. Over the years, they have become more reliable as well as computer-efficient. The mask R-CNN is the latest implementation produced by Facebook researchers and is an excellent basis for server side object detection models.

The key difference between object detection and classification algorithms is that we have to locate the bounding box of an object in the detection algorithm and we try to classify the object in the classification algorithm. In situations where you need to draw each object on the screen on one slide, remember that you do not usually need to fill in the box of each object. As you do not know how many objects there will be, do not worry about that.

Clustering algorithms break up the picture into thousands of parts, scanning through the parts to find regions that are in the picture. Ross Girshick suggested a method to break this up into different categories, which he called spatial proposals, and showed that the regional proposals could be evaluated using the $k = 2$ technique. But now you (you'll need to) identify a minimal number (only 2,000 regions) of regions, instead of having to identify a vast number of regions. The proposals for the 2,000 regions were generated with the powerful, yet easy to use, algorithm.

These 2,000 proposals for candidate regions are twisted in a square in a FNN. The functional vector is computed by the CNN model on the top of a capture-recurrent encoder. The CNN is the feature extractor that forms a thick outcome layer, consisting of the features extracted from the image. Each feature is then translated into a classifier to tell the computer the candidate region for a candidate area. The algorithm on which the system is based has many more benefits. It will be able to predict the location of an object and give accurate values for the bounding box, or region of analysis. For example, the algorithm would have predicted the appearance of a person within this region, but because of the halved face image embedded within, the person's appearance had been cut in half in this region proposal. Since the offset values change the bounding box, the change in the closed area will help to limit the area request.

It takes quite a lot of effort to train the information retrieval network as 2,000 regional proposals have to be classified by content. It cannot be possible in the real time due to the fact that each test image requires about 47 seconds. The algorithm is fixed for all searches that will be conducted. At that stage, none of the learning approaches has occurred. By making poor candidates, the person could make more proposals to make other poor candidates.

Along with the researchers who found R-CNN, the same author of the previous chapter found Fast R-CNN. It is a faster algorithm for object detection. It improves the efficiency of object detection and can help a lot. However, rather than feeding the CNN input picture directly to CNN, we are feeding CNN the CNN regional suggestions structure to produce a fully convolutional CNN. We created a feature map of the microtask proposal area. As the proposals for microtasks are divided into squares, we fixed the scale of each of those squares using a kernel pooling layer so that it could be fed into a fully connected layer. We use a softmax layer from the ROI function vector to determine the class of the area; the bounding box offset values are added to the "box" attribute and used to build the "layout" object.

The reason "Fast R-CNN" is faster than R-CNN is because it requires 2,000 dimension vector arrays to be generated instead of 2,000 feature submissions, resulting in a more accurate result. As opposed to convolution, the deconvolution is carried out only once per image, producing a feature map.

The area proposal picking algorithm (Quick R-CNN) and region proposal picking algorithm (R-CNN) use a selective searching method. Selective search is a time-consuming and long method that slows down network connections. As a result, Shaoqing Ren et al. have developed an object detection algorithm that eliminates the selective search algorithm and enables the network to learn about the region suggestions.

Like Fast R-CNN, the picture is an input for a convolutional network. The network uses the picture as input and creates a feature map as output. The region proposal features are extracted from a separate network that ranks the region proposals instead of interrogating the region proposals through the specific features in the function map. After determining the area of the circle and establishing its bounding box, the proposed area is reshaped to make it into the area that we ultimately want to identify and outline.

Mask R-CNN is conceptually simple: Mask R-CNN, which uses R-CNN (REinforcement-Learning CNN) with masking, projects each set of bounding boxes as class labels and masks those labels onto the correct objects. In addition, we will add an extra branch on the right to which the mask will take. The R-CNN mask is therefore a typical thing and it is obvious. The ear-shaped portion of the mask does not allow the removal of dense matter from the internal edge, but rather, it helps accentuate and enlarge the object's outer edge facing spatial structure. Out of all of the above-mentioned components, the most important is the R-CNN that supports the aligned views for the images, but the other components that do not include

R-CNN are still very important. Algorithms, such as R-CNN and You Only Look Once (YOLO) (Chen et al., 2020, p. 806), are extremely helpful to recognise objects very quickly. These cameras try to detect objects in one shot. Fast processing with great accuracy can be expected from the lab.

11.2.2   YOLO, MobileNet + SSD, SqueezeDet

There are many different kinds of noise-suppressing devices, including vests. The other principal distinction is that the negative is better than the positive. The MobileNet + SSD models (Li et al., 2018, p. 1678) involve an encoder based solely on MobileNet and SqueezeNet. In YOLO, the use of a single convolutional network paired with a set of rules that happens as the model runs (YOLO) predicts many class possibilities at once.

Previously, we used the region of the picture to identify the object in the scene. The output may not be perfect. Images are the parts of an image that contain the objects that make it up. YOLO is a very different algorithm that is akin to HOG/SSIM. In YOLO, the likelihood of belonging and class probabilities are determined with a single convolutional RNN.

YOLO trains directly on images and detects criminals successfully and swiftly. Based on this unified paradigm, we have a number of advantages over a conventional approach. YOLO is a very fast real-time multi-object detection algorithm. You don't need a complicated pipeline because it's going to have a regression problem on frame detection. Because it is a new technology, we only use our neural network for predictions on new experiments. Our current Fast Ring builds are running at ~45 FPS without the assistance of a Titan X card to speed things along.

NVIDIA has a version that can display over 150 FPS. In a little less than 25 milliseconds, we can definitely complete the delay. YOLO technology gives us a much higher accuracy in real time.

11.2.3   CSRNet-Congested Scene Recognition

The technical paper published by Li, Zhang and Chen from the University of Illinois in April 2018 reduced the problem by drawing a network density map, which provided accurate and extensive information. Unfortunately, accurate distribution patterns are difficult to establish. The major difficulty is because the density values are randomly generated pixel by pixel. So, we have to provide spatial coherence so that their uniformity will be smooth although it is extremely difficult because of the diverse scenes, different crowd clusters and different camera perspectives, apart from traditional methods (DNNs). The recent developments in DNN-based methods are proving useful in semantic segmentation tasks, which signify progress in visual saliency. Unlike the study made by Chen et al. (2020, p. 211) and Ma et al. (2019, p. 100) that utilises deep CNN for auxiliary, here the focus is on the design of a CNN-based density map generator. We use pure convolutional layers as a base of supporting input images. We opt for smaller sized convolution filters throughout the network. Here, the features are trained from VGG-16 and three dilated convolutional layers are used (since pooling layers are not used).

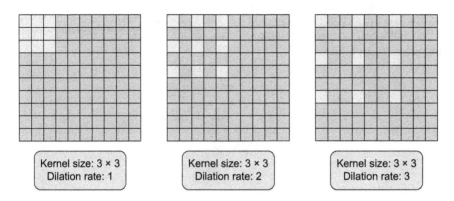

FIGURE 11.4 3×3 Conventional kernels with different dilation rates.

Instead of increasing the kernels parameters, we can expand them. So, here we are inverting the entire image of the kernel. When the rate of dilatation is increased to 2, the kernel enlarges into that shape (follow the labels below each image). Splitting the layers is an alternative to aggregation. Figure 11.4 shows the 3 X 3 conventional kernels with different dilation rates.

11.2.3.1   Mathematics in the CSRNet

The following mathematics are from the CSRNet research paper (Sabokrou et al., 2018, p. 94).

Suppose we have an input x(m,n), a filter w(i,j) and the dilation rate r. The output y(m,n) will be as follows:

$$y(m,n) = seriesSum \ (seriesSum(x(m + r \cdot i, n + r \cdot j)w(i,j), j, 1, N), i, 1, M)$$

This equation can be generalised using a (k*k) kernel with a dilation rate of one (x). From this kernel, it comes to the following:

$$\left([k + k - 1] * [k + k - 1] * [k + k - 1] * [k + k - 1] \right)$$

And we have generated the ground truth of all the images. Each head in the image is blurred using a Gaussian kernel. Each image is cropped into nine patches, and its dimensions are one-fourth of the original size of the image.

The first four patches are re-cropped every quarter and the other five are re-cropped randomly when the picture is taken. Finally, each patch is replicated to double the training set.

These are the architecture particulars of CSRNet in a nutshell. The stochastic gradient descent is an algorithm used to train the neural network of an end-to-end structure. The training is achieved at a rate of 10e−6. We chose to use the Euclidean distance method in our spatial regression. It is understood as follows:

$$L(\Theta) = \frac{1}{2N} \sum_{i=1}^{N} \left\| Z(X_i; \Theta) - Z_i^{GT} \right\|_2^2$$

where N is the size of the training batch. The evaluation metric used in CSRNet is MAE and MSE, i.e., mean absolute error and mean square error, respectively. These are given by the following:

$$MAE = \frac{1}{N}\sum_{i=1}^{N}\left|C_i - C_i^{GT}\right|$$

$$MSE = \sqrt{\frac{1}{N}\sum_{i=1}^{N}\left|C_i - C_i^{GT}\right|^2}$$

Here, C_i is the estimated count:

$$C_i \sum_{l=1}^{L}\sum_{w=1}^{W}zl, w$$

L and W are the length and width of the predicted density map, respectively.

The model will first predict the distribution of dense pixels in an image. If no person is present, the pixel value will be zero. A price will be assigned to every person corresponding to its average value. So, measuring the total pixel values corresponding to an individual will give the exact count of people in that picture.

11.3   IMPLEMENTATION

This section explores a visual as well as theoretical representation of the proposed model of the application server and client-side designs and architectures. Using video and image analyses and model training system, our software determines the current rush in trains and also using previous train rush history, the software will give accurate results of current crowd density in trains. So based on the results, users can take an appropriate train. This application also helps the government to make a decision whether extra trains are needed or not. The public transport chosen is a local train. The system architecture of crowd detection is shown in Figure 11.5. Following are the constraints for these system implementations:

- To set up a CCTV camera on a railway station is legally not possible. Hence, the data that will be used is from videos taken from mobile and YouTube and these videos are past videos.
- Complex nature of CNN for implementation.
- Good connectivity to the Internet.

Live video recorder: This module will record video at regular interval and send that to the server.

User application: An Android/iOS app that will show the users the upcoming train and the crowd density in that train.

Server: A server will be a system in which actually machine learning script will run, which goes to detect the density of crowd. An app server will invoke an ML

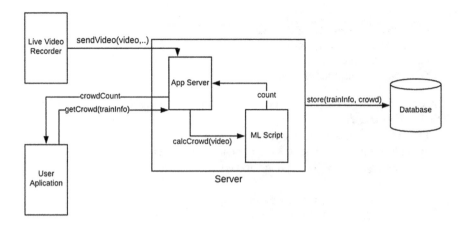

FIGURE 11.5 System architecture of crowd detection.

script to calculate the count of humans in that particular frame captured by a live video recorder module. A server also communicates with database and it will store information regarding train and crowd count in database. The server will also serve the request coming from user application and send crowd count to user.

Database: A database will store the following information such as train number, starting time, destination, last station, crowd count and timestamp.

App server: Serves the user app with data of crowd.

ML script: A Python script is going to calculate the crowd density by analysing the video feeds.

Flow: A live video recorder will record the videos of 35s from when the train arrives and send the video to server in an HTTPs POST request, which is going to carry the following data in JSON format:

```
{
"trainNo": "12345",
"stationName": "Dahisar";
"timestamp": "27-11-2019 12:24:32";
"video": file;
}
```

Now the server will send the video to the ML script, which will be a Python script that will execute as a separate process on the server spawned by the app server; this ML script will do the processing and return the crowd count to the video.

Once an app server has the count, it will execute an algorithm and store the count of a train in the database along with the train information.

11.3.1 ALGORITHM

The algorithm mentioned before is as follows:

```
alreadyPT = max(0,10)
up = alreadyPT - Min
```

```
down = Max - alreadyPT
finalTN = alreadyTN - down + up
finalTN = alreadyTN + 2*alreadyPT - Max - Min
alreadyPT: total count of passengers already in train
```
finalTN: total number of passengers

Once we have the data, we will push that to the user application, which will render the data on the user app.

The videos recorded for each station are being uploaded to a website by the admin along with the train id. The website then sends those videos to the servers and the rest processing of the videos is done at the server end and the count of people in the train is updated to the database.

The server is node based that runs on an EC2 instance on Amazon Web Services (AWS). The server does the following tasks:

1. It serves the request of the client for the crowd count live video recorder.
2. It serves the web page of a live video recorder to admin.
3. It takes the video uploaded to it and spawns a Python process and passes the video location. Now the child process calculates the crowd count and returns to the server.
4. Once the crowd count is returned, the server writes that crowd count to database.

An NoSQL-MongoDB hosted on MongoDB Atlas, which is backed by AWS, is used for database.

The client app will request the server to send the crowd count for each train and will display the crowd density to the users. It is an offline application.

The implementation of model was done with the help of two open-source project in our project (Chen et al., 2020, p. 211; Ma et al., 2019, p. 100).

11.3.1.1 Building of Data Set

We have used the ShanghaiTech data set for the task of crowd counting. There is a MATLAB file (.mat) that contains the names of the images and their locations available to the data analyst who wants to use those images for training purpose. First of all, the data is divided into two portions. Part A involves a higher number of people. Part B depicts open sparse crowd scenes.

11.3.1.2 Data Preprocessing

When we talk about data, we usually think of many huge data sets. However, as of today, some data is also being stored in the form of images, audio and video materials.

Machines cannot understand free written words, images or videos as it is. It won't be enough if we just present a static set of images and expect our machine learning system to be automatically trained by that. Data sets are basically a collection of data objects.

Data objects are described by a number of attributes such as the mass of a physical object or the time at which an event occurs. Attributes can be referred to as variables, features, characteristics or dimensions.

One of the main purposes of the preprocessing stage is to convert the ground truth into density maps. For each captured video, the image data set provided a sparse matrix consisting of the face annotations of that video. The sparse matrix was converted into a two-dimensional (2D) histogram by passing through a low-frequency Gaussian filter. The sum of all the cells in the density map results in the overall population of the space.

11.3.1.3 Algorithm for Creating Density Map for Given Head Annotation
We will be converting an image to a density map.

1. Build a tree of head annotations. This tree is known as k-d tree. k-d Tree is one of the types of data structure, which allows fast computation for K-nearest neighbors.
2. Find the average distances for each head with K (in this case 4)-nearest heads in the head annotations. Multiply this value by a factor, 0.1, as suggested by the author of the chapter.
3. Put this value as sigma and convolve using the 2D Gaussian filter.

11.3.1.4 Model Building
Data science team needs to set up data sets for training, testing and production phases. These data sets enable data scientists to develop statistical models and test them by throwing aside certain data sets to see if the predictions hold.

Team provides data sets for testing, training and production. In addition, in this phase, the team builds models based on the models created in the previous phase. The team also will have to make sure that the models can be executed on the system and run smoothly.

The CSRNet model uses CNNs that take the input image and give the resultant density image. This model does not make any use of any fully connected layers and thus can consist of input images of variable sizes. The model is a learning system based on a large amount of data and it generates data without losing any information. It's not necessary to redesign/reshape the image while inferring. The model architecture is such that a density image of size (x/8,y/8,1) is produced by describing a 2D input in the target space.

Figure 11.6 shows a density map where concentration of points is shown.

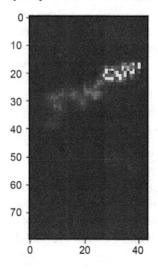

FIGURE 11.6 Density map.

The architecture of this application is divided into two parts: front end and back end. The back end of the VGG16 pre-trained model consists of 13 pre-trained layers (10 convolution layers and 3 max pooling layers). This is not taken into account. The second layer is dilated convolutional layer 2 as per the suggestions from CSRNet.

Batch normalisation support has been implemented in the code. We prepared a custom VGG16 model and reprocessed the pre-trained layers to this model.

11.3.1.5   Variable Size Input

Keras (n.d.-b) is made for humans and not computers. Keras provides clean APIs with no magic and is a very easy-to-use library. It has a large number of plugins and extensions. In Keras, it is difficult to train a model that depends on the size of the input. Keras does not allow different size inputs to be trained together in the same training run. Another method is to combine all images and input them as a batch. It does not contain many images of the same size in the ShanghaiTech dataset. These batches could not be started. Another method is to train the character independently and run a preprocessing loop over each character. This is hardly optimal in terms of memory and computation. We managed to complete our training process using custom data generator in Keras. Training with a data generator saves on memory usage and time.

The report specifies cropping photographs as a part of data manipulation. PyTorch is an open-source machine learning library first developed by the Facebook AI Research Lab. It is free software released under modified BSD license. However, the PyTorch implementation does not crop the images. With these changes, we have also added the functionality for cropping. This model has been trained to learn without cropping the images.

Both models trained using the same architecture on 200 epochs. The other hyper-parameters used in the CSRNet paper and PyTorch (Chen et al., 2020, p. 211) implementation were kept intact.

11.3.1.6   Testing and Working

A key purpose of model evaluation is to estimate the generalisation accuracy of a model on unseen (out-of-sample) data. Here, we have created a splitter function: splitter () that takes video as input and splits them into image as per 0.5/1 frame per seconds. Then we feed these images to the model build and generate a density map and obtain the crowd per image. The Output of splitter function with a duration of 1 second is shown in Figure 11.7.

We then extract the min, max and platform value variable and send to NodeJS server for further processing.

11.3.1.7   Variable Working

- Max: It is a maximum amount of crowd found from the images feed in.
- Min: It is minimum amount of crowd found from the images feed in.
- Platform value: It is a maximum amount of crowd found from the images feed in before train arrival (assumption taken images till 10 seconds are used to extract this value).

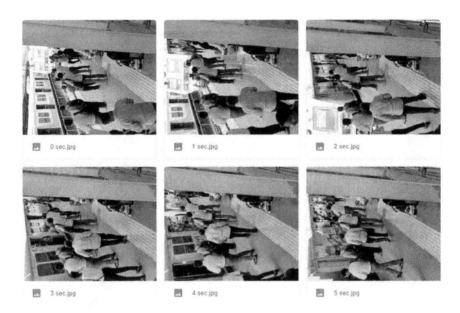

FIGURE 11.7   Output of splitter function with a duration of 1 second.

11.3.1.8   Testing

The testing is done based on how the model performs based on a different video quality of the same video. Here, we have taken two videos from the YouTube as data.

The first video is being tested on 360p and 720p. The second video is being tested on 360p, 720p and 1080p.

In both the test videos, the value of the variable finalTN, i.e., count of people in train, was considered 50.

11.3.1.9   Video 2

From the previous two videos, it can be concluded that the model gives a good accuracy even when the data is of 360p.

From the result of video 1, the value got from 360p quality is 89.15% accurate compared to the value got from 720p.

From the result of video 2, the value got from 360p quality is 72.72% accurate compared to the value got from 720p. The results of 720p and 1080p are almost similar. The value got from 720p is 97.77% accurate compared to the value got from 1080p. The results of video 1 and video 2 are shown in Figure 11.8 to Figure 11.12.

This system presents an application that helps users in seeing the crowd density in the local train. This was a prototype model, and seeing the results, it can be concluded that such a system can be implemented. The application was made only for androids. But it can also be made for iOS. The method used for finding crowd density in local trains can also be implemented for other transport systems such as buses and metros. The core concept remains the same.

360P Video1

```
[ ]  # Variable  which is to send for requirements.
     # Important variable for Application
     min3=min_value
     max3=max_value
     plat3=platform_max

     print("min value : ",min3)

     print("max value :",max3)

     print("platform value :",plat3)

     finalTN=AlreadyTN+2*plat3-max3-min3

     print("So the final crowd value in Train will be")
     print(round(finalTN))
```

```
min value :   26.562962
max value : 50.631294
platform value : 50.631294
So the final crowd value in Train will be
74.0
```

FIGURE 11.8   Result of video 1 which is of quality 360p.

720P-Video-1

```
[ ]  # Variable  which is to send for requirements.
     # Important variable for Application
     min2=min_value
     max2=max_value
     plat2=platform_max

     print("min value : ",min2)

     print("max value :",max2)

     print("platform value :",plat2)

     finalTN=AlreadyTN+2*plat2-max2-min2

     print("So the final crowd value in Train will be")
     print(round(finalTN))
```

```
min value :   43.538475
max value : 76.998985
platform value : 76.998985
So the final crowd value in Train will be
83.0
```

FIGURE 11.9   Result of video 1 which is of quality 720p.

360P-Video-2

```
[ ]  # Variable  which is to send for requirements.
     # Important variable for Application
     min3=min_value
     max3=max_value
     plat3=platform_max

     print("min value : ",min3)

     print("max value :",max3)

     print("platform value :",plat3)

     finalTN=AlreadyTN+2*plat3-max3-min3

     print("So the final crowd value in Train will be")
     print(round(finalTN))

     min value :   25.445312
     max value : 39.80584
     platform value : 35.49704
     So the final crowd value in Train will be
     56.0
```

FIGURE 11.10   Result of video 2 which is of quality 360p.

720P-Video-2

```
[ ]  # Variable  which is to send for requirements.
     # Important variable for Application
     min2=min_value
     max2=max_value
     plat2=platform_max

     print("min value : ",min2)

     print("max value :",max2)

     print("platform value :",plat2)

     finalTN=AlreadyTN+2*plat2-max2-min2

     print("So the final crowd value in Train will be")
     print(round(finalTN))

     min value :   36.057423
     max value : 50.487923
     platform value : 40.37197
     So the final crowd value in Train will be
     44.0
```

FIGURE 11.11   Result of video 2 which is of quality 720p.

1080P--Video--2

```
[ ]  # Variable  which is to send for requirements.
     # Important variable for Application
     min1=min_value
     max1=max_value
     plat1=platform_max
     # print(end-start)
     print("min value : ",min1)

     print("max value :",max1)

     print("platform value :",plat1)


     finalTN=AlreadyTN+(2*plat1)-max1-min1

     print("So the final crowd value in Train will be")
     print(round(finalTN))
```

```
min value :   53.537926
max value :   70.03238
platform value :  59.49968
So the final crowd value in Train will be
45.0
```

FIGURE 11.12　Result of video 2 which is of quality 1080p.

11.4　CONCLUSION AND FUTURE SCOPE

Here, we studied about object detection techniques that can be used to find out crowd estimation at public gathering. Here, we considered crowd estimation in train through object detection. Here, we have also done testing on YouTube video data and obtained adequate results. This application can be used for places where we need to estimate crowd and, according to that, we can make additional provisions to accommodate such crowd as in bus and train services. Here, we worked with few algorithms but, in future, more and more algorithms need to be tested and applied for the same purpose. So an accuracy of crowd estimation can be improved.

REFERENCES

Chen, J., Su, W., & Wang, Z. (2020). Crowd counting with crowd attention convolutional neural network. *Neurocomputing*, *382*, 210–220. https://doi.org/10.1016/j.neucom.2019.11.064

Chen, W., Huang, H., Peng, S., Zhou, C., & Zhang, C. (2020). YOLO-face: a real-time face detector. *The Visual Computer*, *37*(4), 805–813. https://doi.org/10.1007/s00371-020-01831-7

Forsyth, D., & Ponce, J. (2011). *Computer Vision: A Modern Approach* (2nd ed.). Pearson.

Girshick, R., Donahue, J., Darrell, T., & Malik, J. (2016). Region-based convolutional networks for accurate object detection and segmentation. *IEEE Transactions on Pattern Analysis and Machine Intelligence*, *38*(1), 142–158. https://doi.org/10.1109/tpami.2015.2437384

Kusters, A. (2017). When transport becomes a destination: deaf spaces and networks on the Mumbai suburban

Li, Y., Huang, H., Xie, Q., Yao, L., & Chen, Q. (2018). Research on a surface defect detection algorithm based on MobileNet-SSD. *Applied Sciences*, *8*(9), 1678. https://doi.org/10.3390/app8091678

Ma, J., Dai, Y., & Tan, Y. P. (2019). Atrous convolutions spatial pyramid network for crowd counting and density est

Sabokrou, M., Fayyaz, M., Fathy, M., Moayed, Z., & Klette, R. (2018). Deep-anomaly: fully convolutional neural network for fast anomaly detection in crowded scenes. *Computer Vision and Image Understanding*, *172*, 88–97. https://doi.org/10.1016/j.cviu.2018.02.006

ShanghaiTech_Crowd_Counting_Dataset.zip. (n.d.). Google Docs. https://drive.google.com/file/d/16dhJn7k4FWVwByRsQAEpl9lwjuV03jVI/view

Viola, P., & Jones, M. (2001, February). *Robust Real-Time Object Detection*. https://www.hpl.hp.com/techreports/Compaq-DEC/CRL-2001-1.pdf

12 Analysis of a Machine Learning Algorithm to Predict Wine Quality

Nilesh Bhikaji Korade

CONTENTS

12.1 INTRODUCTION

To increase the quality of product, testing is an important phase to ensure the quality. Today, all types of industries are adopting and applying new technologies to ensure test quality of product [1]. Testing quality of products by human expertise is an expensive and time-consuming process [3]. This chapter explores different machine learning techniques, which can be used to determine wine quality such as logistic regression, decision tree, random forest, support vector machine (SVM), AdaBoost classifier, and gradient boosting classifier. These techniques perform quality assurance processes with the help of available characteristics of product and automate the process by minimising human interference. The chapter also identifies the important

DOI: 10.1201/9781003122401-12

features to predict the values of dependent variables. Wine quality assessment is one of the key elements that can be used for certification that helps to assure wine quality in the market. The input variables in the Red Wine Dataset are fixed acidity, volatile acidity, citric acid, residual sugar chlorides, free sulfur dioxide, total sulfur dioxide, density, pH, sulfates, and alcohol. The quality is in the range of 1–10, the higher value indicates a better quality of wine [8].

The chapter is organised as follows: Section 12.2 provides the description of related work in this area. Section 12.3 discusses the proposed methodology, the dataset used, and description of machine learning algorithm in detail. Experimental results and analysis are explained in Section 12.4. Conclusion is drawn in Section 12.5.

12.1.1 OBJECTIVES

The first objective is to experiment with different classification methods to see which yields the highest accuracy. The second objective is to determine which features are the most indicators of good quality wine.

12.2 RELATED WORK

Most of the researchers used machine learning algorithms to assess wine quality, but still a huge scope is available for improvement. Yogesh Gupta [1] in his article explores the usage of linear regression to determine important features for prediction and usage of neural network and SVM in predicting the values. This chapter explores the usage of machine learning techniques in two ways: first, how linear regression determines important features for prediction and, second, the usage of neural network and SVM in predicting the values. Terence Shin [2] used Kaggle's Red Wine Quality Dataset to build various classification models to predict whether a particular red wine is "good quality" or not. Devika Pawar et al. [3] used several techniques such as logistic regression, stochastic gradient descent, support vector classifier, and random forest to predict the wine quality. The analysis shows that quality increases as residual sugar is moderate and does not change drastically, so this feature is not essential as compared to others such as alcohol and citric acid. Saini et al. [4] performed fundamental and technical analysis of various algorithms used for predicting future stock market prices and found long short-term memory neural network (LSTM NN) producing better results as compared to other techniques. The study of Yunhui Zeng et al. [5] take wine quality evaluation as the research object, establishes the analysis and evaluation model of wine quality, and explores the influence of physical with chemical indicators of grapes and wine on the wine quality. This study performed multiple linear regression analysis and finds that there is a positive correlation linear relationship between the scores of the aroma of wine quality and C_2H_6O, $C_6H_{12}O_2$, etc. Jambhulkar et al. [6] used various techniques to predict heart disease using wireless sensor network. To predict heart disease, they collected data from Cleveland dataset and extracted important. Zaveri et al. [7] predicted different diseases such as cancer, TB, and diabetes using data mining techniques. Yesim Er et al. [8] evaluated different classifiers like k-nearest neighborhood, random forests, and SVMs on wine datasets. They applied principal component analysis and find that the success rate of quality classification for white wine has decreased for cross-validation mode and for percentage split mode, whereas the success rate of quality classification for red wine has increased for cross-validation mode and for percentage split mode.

12.3 PROPOSED METHODOLOGY

In this study, machine learning algorithms are used to predict wine quality. Figure 12.1 describes the processing in proposed methodology. First, wine dataset is preprocessed. Further, data is divided into training (80%) and testing sets (20%); the training set is used to train model by using logistic regression, decision tree, random forest, SVM, AdaBoost classifier, and gradient boosting classifier algorithm and is used to find accuracy of different model. Then, a conclusion is drawn that the trained model can be used to identify the accuracy of the testing set. The accuracy of different algorithms is evaluated and compared to find the best algorithm to predict quality of wine.

12.3.1 DATASET

In this study, the collection of Red Wine Dataset [9] is used. There are total *1599* samples for red wine. Each sample consists of 12 physiochemical variables: fixed acidity, volatile acidity, citric acid, residual sugar, chlorides, free sulfur dioxide, total sulfur dioxide, density, pH, sulfates, alcohol, and quality rating. The quality scale ranges from 0 to 10 where 0 is very bad and 10 is very excellent. Due to some deficiencies in dataset, it is not possible to use wine collections without preprocessing. One of the major deficiencies is the large amplitude of variable values, e.g., sulfates (0.3–2) vs. sulfur dioxide (1–72), there are some missing values. The missing values are filled by taking mean. The inconsistency in dataset affects predictions due to influence making by some variables, such inconsistency is solved by linear transformation by dividing all the input values

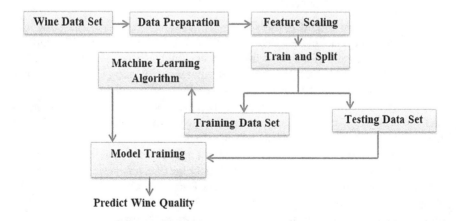

FIGURE 12.1 Proposed system architecture.

by maximum variable values. The wine quality is converted to a binary output, "1" means good quality wine (having a score of 7 or higher) and "0" mean bad quality wine (having a score below 7).

12.3.2 Machine Learning Technique

The description of different machine learning algorithms used in this study is given in the following sections.

12.3.2.1 Logistic Regression

Does a machine learning algorithm come under the supervised learning technique? It is used for predicting the categorical-dependent variable using a given set of independent variables. Logistic regression predicts the output of a categorical-dependent variable and the outcome must be a categorical value. It can be 0 or 1, Yes or No, True or False, etc., but instead of giving the exact value as 0 and 1, it gives the probabilistic values that lie between 0 and 1. We know the equation of the straightline can be written as:

$$y = b_0 + b_1 x_1 + b_2 x_2 + \cdots b_n x_n$$

In logistic regression, y can be between 0 and 1 only, so for this let's divide the above-mentioned equation by $(1 - y)$.

$$y/(1-y); 0 \text{ for } y = 0 \text{ and infinity for } y = 1$$

But we need range between −[infinity] to +[infinity], then take logarithm of the equation and it will become.

$$\log \frac{y}{1-y} = b_0 + b_1 x_1 + b_2 x_2 + \cdots b_n x_n.$$

This equation is the final equation for logistic regression.

12.3.2.2 Support Vector Machine

Can a supervised machine learning algorithm be used for both classification and regression challenges? However, it is mostly used in classification problems. SVM constructs a hyperplane in a high- or infinite-dimensional space, which can be used for classification, regression, or other tasks.

The objective of the SVM algorithm is to create the best line or decision boundary that can segregate n-dimensional space into classes so that we can easily put the new data point in the correct category in the future. This best decision boundary is called a hyperplane. Hyperplane can be multiple lines or decision boundaries to segregate the classes in n-dimensional space, but to classify the data points we need to find out the best decision boundary. This best boundary is known as the hyperplane of SVM. Support vectors are the data points or vectors that are the closest to the hyperplane and affect the position of the hyperplane. SVM chooses the extreme points or vectors

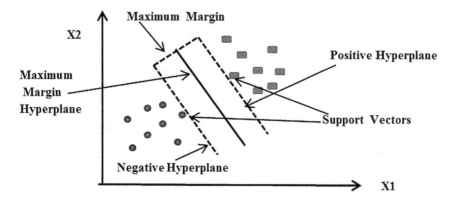

FIGURE 12.2 Support vector machine classifier.

that help in creating the hyperplane. These extreme cases are called support vectors, and hence algorithm is termed SVM. Consider Figure 12.2 in which there are two different categories that are classified using a decision boundary or hyperplane, called positive and negative hyperplanes.

12.3.2.3 Decision Tree

Can a supervised learning technique be used for both classification and regression problems? However, it is mostly used for solving classification problems. It is a tree-structured classifier, where each leaf node represents the outcome, internal nodes represent the features of a dataset, and branches represent the decision rules. As compared to other algorithms, decision trees require less effort for data preparation during preprocessing. A decision tree does not require normalisation of data and scaling of data as well. Missing values in the data also do not affect the process of building a decision tree to any considerable extent. But a small change in the data can lead to a large change in the structure of the optimal decision tree. Calculations can become very complex, particularly if many values are uncertain and/or if many outcomes are linked. The main issue while implementing a decision tree is that how to select the best attribute for the root node and for sub-nodes. This technique attributes selection measure or ASM technique is used to solve such problems. There are two popular techniques for ASM information gain and Gini index.

$$\text{Information gain} = \text{Entropy (S)} - [(\text{weighted avg}) \times \text{entropy (each feature)}]$$

Entropy is a metric to measure the impurity in a given attribute. It specifies randomness in data.

$$\text{Gini index} = 1 - \sum_{j} P_j^2$$

Gini index is a measure of impurity or purity used while creating a decision tree in the classification and regression tree (CART) algorithm.

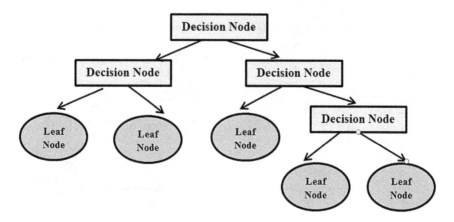

FIGURE 12.3 Decision tree classifier.

The decision tree is shown in Figure 12.3 where internal nodes represent the features of a dataset, and branches represent the decision rules.

12.3.2.4 Random Forest

Can a supervised learning algorithm be used for both classification and regression problems? The "forest" it builds is an ensemble of decision trees, usually trained with the "bagging" method. The bagging method is a combination of learning models and increases the overall result. Random forest is a learning method that operates by constructing multiple decision trees. The final decision is made based on the majority of the trees chosen by the random forest. As shown in Figure 12.4, random forest creates n number of decision trees by randomly selecting records from dataset. Instead of relying on one decision tree, the random forest takes the prediction from each tree and based on the majority votes of predictions, and it predicts the final output. The greater number of trees in the forest leads to higher accuracy and prevents the problem of overfitting.

There are two stages in random forest algorithm, first is to create random forest and second is to make a prediction from the random forest classifier created in the first stage.

1. Randomly select "K" features from total "m" features where K ≪ m.
2. Among the "K" features, calculate the node "d" using the best split point.
3. Split the node into daughter nodes using the best split.
4. Repeat 1–3 steps until "l" number of nodes has been reached.
5. Build forest by repeating steps 1–4 for "n" number times to create "n" number of trees.

In the next stage, with the random forest classifier created, we will make the prediction.

1. Take the test features and use the rules of each randomly created decision tree to predict the outcome and store the predicted outcome.

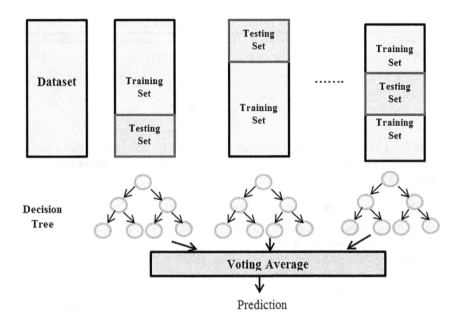

FIGURE 12.4 Random forest algorithm classifier.

2. Calculate the votes for each predicted target.
3. Consider the high voted predicted target as the final prediction from the random forest algorithm.

12.3.2.5   AdaBoost Classifiers

Is it best to boost the performance of decision trees on binary classification problem? AdaBoost algorithm, Adaptive Boosting, is a boosting technique that is used as an ensemble method in machine learning. The weights are reassigned to each instance, to incorrectly classified instances. Boosting is used to reduce bias as well as the variance for supervised learning. It works on the principle where learners are grown sequentially.

AdaBoost refers to a particular method of training a boosted classifier. A boost classifier is a classifier in the form

$$F_T(X) = \sum_{t=0}^{T} f_t(x)$$

where f_t is a weak learner that takes an object x as input and returns a value indicating the class of the object. For example, in the two-class problem, the sign of the weak learner output identifies the predicted object class and the absolute value gives the confidence in that classification. The Tth classifier is positive if the sample is in a positive class and negative otherwise.

An output hypothesis h(x$_i$) is produced by each weak learner for each sample in the training set. At each iteration t, a weak learner is selected and assigned a coefficient α_t such that the sum training error E_t of the resulting t-stage boost classifier is minimised.

$$E_t = \sum E\left[F_{t-1}(x_i) + \alpha_t h(x_i)\right]$$

where $F_{t-1}(x_i)$ is the boosted classifier that has been built up to the previous stage of training. E(F) is some error function, $f_t(x) = \alpha th(x_i)$ is the weak learner that is being considered for addition to the final classifier.

12.3.2.6   Gradient Boosting Classifiers

Are gradient boosting classifiers a group of machine learning algorithms? It creates a strong predictive model by combining many weak learning models together. Decision trees are usually used when doing gradient boosting. Gradient boosting classifiers are the AdaBoosting method combined with weighted minimisation, after which the classifiers and weighted inputs are recalculated. The main objective of gradient boosting classifiers is to minimise the loss or distinguish between the actual class value of the training example and the predicted class value. The gradient boosting classifier depends on a loss function. A custom loss function can be used, and many standardised loss functions are supported by gradient boosting classifiers, but the loss function has to be differentiable. Gradient boosting systems don't have to derive a new loss function as every time the boosting algorithm is added, rather any differentiable loss function can be applied to the system. Classification algorithms frequently use logarithmic loss, while regression algorithms can use squared errors. Gradient boosting systems have two other necessary parts, a weak learner and an additive component. Gradient boosting systems use decision trees as their weak learners. Regression trees are used for the weak learners, and these regression trees output real values. Because the outputs are real values, as new learners are added into the model and the output of the regression trees can be added together to correct for errors in the predictions. A procedure similar to gradient descent is used to minimise the error between given parameters. This is done by taking the calculated loss and performing gradient descent to reduce that loss.

Gradient boosting classifiers algorithm:
Input training set {x$_i$, y$_i$} where i = 1...n, a differentiable loss function L(y, F(x)), and M is number of iteration.

Step 1: Initialise model with constant value $F_0(x) = \arg \min \sum_{i=1}^{n} L\,(y_i, \gamma)$

Step 2: For m = 1 – M

1. Compute so-called pseudo residuals

$$r_{im} = -\left[\frac{\delta L\left(y_i, F(x_i)\right)}{\delta F(x_i)}\right] F(x) = F_{m-1(x)} \text{ for } i = 1...n.$$

2. Fit a base learner (or weak learner, e.g., tree) $h_m(x)$ to pseudo-residuals, i.e., train it using the training set$\{x_i, y_i\}$ where i = 1...n.
3. Compute multiplier γ_m by solving the following one-dimensional optimisation problem

$$\gamma_m = \arg \min \sum_{i=1}^{n} L\left(y_i, F_{m-1}(x_i) + \gamma h_m(x_i)\right)$$

4. Update the model

$$F_m(x) = F_{m-1}(x_i) + \gamma_m h_m(x)$$

Step 3: Output $F_M(x)$

12.4   EXPERIMENTAL RESULTS AND ANALYSIS

There are a total of *12* variables in red wine collections as discussed in the previous section. The variable quality rating is considered as dependent variable and the other *11* variables are assumed as predictors or independent variables in this work. The distribution of the quality variable is shown in Figure 12.5. The accuracy of different machine learning algorithms is shown in the below tables.

The performance of the classification models for a given set of test data is drawn by using confusion matrix. It can only be determined if the true values for test data are known. In information retrieval and classification in machine learning, precision is also called positive predictive value that is the fraction of relevant instances among the retrieved instances, while recall is also known as sensitivity that is the

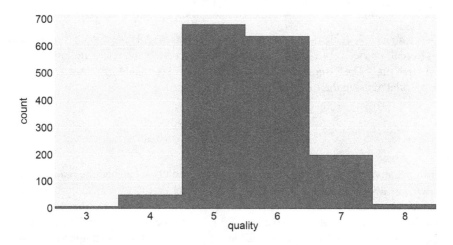

FIGURE 12.5   Distribution of the quality variable.

fraction of relevant instances that were retrieved. Both precision and recall are therefore based on relevance. In statistical hypothesis testing, a type-I error is the rejection of a true null hypothesis also known as a "false-positive" (FP) finding or conclusion; for example, an innocent person is convicted, while a type-II error is the non-rejection of a false null hypothesis also known as a "false-negative" (FN) finding or conclusion; for example, a guilty person is not convicted. The different terms used are described next:

Classification accuracy: It defines how often the model predicts the correct output. It is one of the important parameters to determine the accuracy of the classification problems. The classification accuracy can be calculated as the ratio of the number of correct predictions made by the classifier and all number of predictions made by the classifiers. The formula is given next:

$$Classification\ accuracy = \frac{TP + TN}{TP + TN + FP + FN}$$

Precision: It can be defined as the number of correct outputs provided by the model or out of all positive classes that have predicted correctly by the model, how many of them were actually true. It can be calculated using the below formula:

$$Precision = \frac{TP}{TP + FP}$$

Recall: It is defined as how our model predicted correctly out of total positive classes. The recall must be as high as possible.

$$Recall = \frac{TP}{TP + FN}$$

F-measure: It is difficult to compare models if two models have low precision and high recall or vice versa. The F-score helps us to evaluate the recall and precision at the same time. The F-score is maximum, if the recall is equal to the precision. It can be calculated using the following formula:

$$F-measure = \frac{2 \times Recall \times Precision}{Recal + Precision}$$

True Negative (TN) is the model that has given prediction No, and the real or actual value was also No. True Positive (TP) is the model that has predicted yes, and the actual value was also true. FN is the model that has predicted no, but the actual value was Yes, it is also called type-II error. FP is the model that has predicted Yes, but the actual value was No. It is also called a type-I error. The distribution of quality attribute and their count is shown in Figure 12.5.

TABLE 12.1
Performance Matrix for Logistic Regression Algorithm

	Precision	Recall	F1-Score	Support
0	0.92	0.96	0.94	355
1	0.55	0.38	0.45	45
Accuracy			0.90	400
Macro avg	0.74	0.67	0.69	400
Weighted avg	0.88	0.90	0.89	400

TABLE 12.2
Performance Matrix for Decision Tree Classifier Algorithm

	Precision	Recall	F1-Score	Support
0	0.96	0.92	0.94	355
1	0.53	0.73	0.62	45
Accuracy			0.90	400
Macro avg	0.75	0.83	0.78	400
Weighted avg	0.92	0.90	0.90	400

12.4.1 PERFORMANCE MATRIX AND ACCURACY

The performance matrix and accuracy for logistic regression, decision tree, random forest, SVM, AdaBoost classifier, and gradient boosting classifier are shown next.

Logistic regression gave us an accuracy of 90%. The performance matrix for logistic regression algorithm is shown in Table 12.1.

Decision tree classifier gave us an accuracy of 90%. The performance matrix for decision tree classifier algorithm is shown in Table 12.2.

Random forest classifier gave us an accuracy of 92%. The performance matrix for random forest classifier algorithm is shown in Table 12.3.

SVM gave us an accuracy of 90%. The performance matrix for SVM algorithm is shown in Table 12.4.

TABLE 12.3
Performance Matrix for Random Forest Classifier Algorithm

	Precision	Recall	F1-Score	Support
0	0.95	0.97	0.96	355
1	0.68	0.58	0.63	45
Accuracy			0.92	400
Macro avg	0.82	0.77	0.79	400
Weighted avg	0.92	0.92	0.92	400

TABLE 12.4

Performance Matrix for Support Vector Machine Algorithm

	Precision	Recall	F1-Score	Support
0	0.92	0.97	0.94	355
1	0.58	0.31	0.41	45
Accuracy			0.90	400
Macro avg	0.75	0.64	0.67	400
Weighted avg	0.88	0.90	0.88	400

AdaBoost classifier gave us an accuracy of 89%. The performance matrix for AdaBoost classifier algorithm is shown in Table 12.5.

Gradient boosting classifier gave us an accuracy of 89%. The performance matrix for gradient boosting classifier algorithm is shown in Table 12.6.

12.4.2 COMPARISON OF ACCURACY BY DIFFERENT ALGORITHMS

The different machine learning algorithms like logistic regression, decision tree, random forest, SVM, AdaBoost classifier, and gradient boosting classifier are using physiochemical variables: fixed acidity, volatile acidity, citric acid, residual sugar, chlorides, free sulfur dioxide, total sulfur dioxide, density, pH, sulfates, and alcohol variables. The variable quality rating is considered as dependent variable and the

TABLE 12.5

Performance Matrix for AdaBoost Classifier Algorithm

	Precision	Recall	F1-Score	Support
0	0.94	0.94	0.94	355
1	0.51	0.49	0.50	45
Accuracy			0.89	400
Macro avg	0.72	0.71	0.72	400
Weighted avg	0.89	0.89	0.89	400

TABLE 12.6

Performance Matrix for Gradient Boosting Classifier Algorithm

	Precision	Recall	F1-Score	Support
0	0.94	0.94	0.94	355
1	0.52	0.51	0.52	45
Accuracy			0.89	400
Macro avg	0.73	0.73	0.73	400
Weighted avg	0.89	0.89	0.89	400

TABLE 12.7

Comparison of Accuracy Obtained by Different Algorithms

Sr. No.	Algorithm	Accuracy (%)
1	Logistic regression algorithm	90
2	Decision tree classifier algorithm	90
3	Random forest classifier algorithm	92
4	Support vector machine algorithm	90
5	AdaBoost classifier algorithm	89
6	Gradient boosting classifier algorithm	89

other 11 variables are assumed as predictors or independent variables in this work. Two types of analysis are done in this study firstly, the importance of each algorithm to predict wine quality is identified and secondly, the feature selection is done using best predictors. The accuracy provided by different algorithms is shown in Table 12.7. The logistic regression gives 90% accuracy, decision tree classifier gives 90% accuracy, and so on. The random forest classifier algorithm provides the highest accuracy of 92%.

The random forest algorithm creates n number of decision trees and uses different datasets to train each decision tree. To predict quality of wine, the data is sent to each decision tree to predict some output. The majority of votes are calculated to predict final output. Finally, it can be summarised that random forest classifier algorithm is a better machine learning technique for wine quality predictions on the basis of results. At the same time, more precise predictions can be made by random forest classifier algorithm using selected predictors rather than all predictors.

12.4.3 FEATURE IMPORTANCE

Figure 12.6 shows feature importance based on the random forest model. The 11 physiochemical variables such as fixed acidity, volatile acidity, citric acid, residual sugar, chlorides, free sulfur dioxide, total sulfur dioxide, density, pH, sulfates, alcohol are considered to predict quality of wine. To predict the quality of wine it is not necessary to consider all features, we can remove some of the features while training our model like free sulfur dioxide, pH, residual sugar, fixed acidity, etc. The top three features are alcohol, volatile acidity, and sulfates.

12.4.4 COMPARISON BETWEEN VARIABLES

The comparison between variables for good quality wine and bad quality wine is shown in Tables 12.8 and 12.9. From these figures, we can analyse that good quality wines have higher levels of alcohol, higher levels of sulfates, lower volatile acidity, and higher levels of residual sugar on average. Instead of considering all features, the value of dependent variable can be predicted more accurately if only important features are considered in prediction.

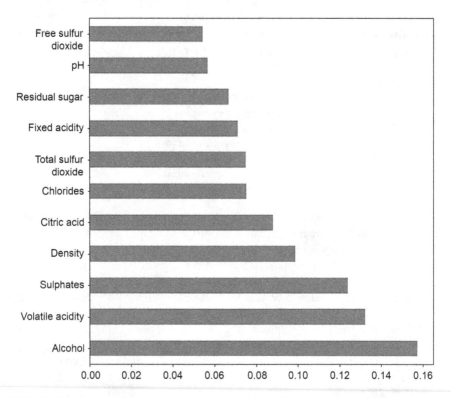

FIGURE 12.6   Feature importance based on the random forest model.

12.5 CONCLUSION

In recent years, the interest in wine industry has been increasing which demands growth in this industry. Therefore, to improve wine production and selling, companies are investing in new technologies. Wine quality certification plays a very important role to sell product in market and it requires wine testing by human experts. This study explores different machine learning techniques to predict the quality of wine. For each classification model, how the results vary whenever test mode is changed is shown in this study. The study includes the analysis of classifiers on red wine datasets. The results are described in percentage of correctly classified instances, precision, recall, F-measure. Different classifiers like logistic regression, decision tree, random forest, SVM, AdaBoost classifier, and gradient boosting classifier are evaluated on datasets. Results from the experiments lead us to conclude that random forest algorithm performs better in classification tasks as compared to other classifiers. The prediction of wine quality achieves maximum accuracy of 92% using random forest algorithm. We can see that good quality wines have higher levels of alcohol on average, higher levels of sulfates on average, a lower volatile acidity on average, and higher levels of residual sugar on average. The study shows that instead of considering all features, the value of dependent variable can be predicted more accurately if only important features are considered in prediction. In future, large dataset can be taken for study and other machine learning techniques may be explored for wine quality prediction.

TABLE 12.8

Comparison between Variables for Good Quality Wine

	Fixed Acidity	Volatile Acidity	Citric Acid	Residual Sugar	Chlorides	Free Sulfur Dioxide	Total Sulfur Dioxide	Density	pH	Sulfates	Alcohol	Quality	Good Quality
Count	217.0	217.0	217.0	217.0	217.0	217.0	217.0	217.0	217.0	217.0	217.0	217.0	217.0
Mean	8.85	0.41	0.38	2.71	0.08	13.98	34.89	1.00	3.29	0.74	11.52	7.08	1.00
Std	2.00	0.14	0.19	1.36	0.03	10.23	32.57	0.00	0.15	0.13	1.00	0.28	0.00
Min	4.90	0.12	0.00	1.20	0.01	3.00	7.00	0.99	2.88	0.39	9.20	7.00	1.00
25%	7.40	0.30	0.30	2.00	0.06	6.00	17.00	0.99	3.20	0.65	10.80	7.00	1.00
50%	8.70	0.37	0.40	2.30	0.07	11.00	27.00	1.00	3.27	0.74	11.60	7.00	1.00
75%	10.10	0.49	0.49	2.70	0.09	18.00	43.00	1.00	3.38	0.82	12.20	7.00	1.00
Max	15.60	0.92	0.92	8.90	0.36	54.00	289.0	1.00	3.78	1.36	14.00	8.00	1.00

TABLE 12.9

Comparison between Variables for Bad Quality Wine

	Fixed Acidity	Volatile Acidity	Citric Acid	Residual Sugar	Chlorides	Free Sulfur Dioxide	Total Sulfur Dioxide	Density	pH	Sulfates	Alcohol	Quality	Good Quality
Count	1382	1382	1382	1382	1382	1382	1382	1382	1382	1382	1382	1382	1382
Mean	8.24	0.55	0.25	2.51	0.09	16.17	48.29	1.00	3.31	0.64	10.25	5.41	0.00
Std	1.68	0.18	0.19	1.42	0.05	10.47	32.59	0.00	0.15	0.17	0.97	0.60	0.00
Min	4.60	0.16	0.00	0.90	0.03	1.00	6.00	0.99	2.74	0.33	8.40	3.00	0.00
25%	7.10	0.42	0.08	1.90	0.07	8.00	23.00	1.00	3.21	0.54	9.50	5.00	0.00
50%	7.80	0.54	0.24	2.20	0.08	14.00	39.50	1.00	3.31	0.60	10.00	5.00	0.00
75%	9.10	0.65	0.40	2.60	0.09	22.00	65.00	1.00	3.41	0.70	10.90	6.00	0.00
Max	15.90	1.58	1.00	15.50	0.61	72.00	165.0	1.00	4.01	2.00	14.90	6.00	0.00

REFERENCES

1. Yogesh Gupta. (2018) "Selection of Important Features and Predicting Wine Quality Using Machine Learning Techniques". *Procedia Computer Science* 125 305–312.
2. Terence Shin. (2020) "Predicting Wine Quality with Several Classification Techniques", May 8, 2020, https://towardsdatascience.com/predicting-wine-quality-with-several-classification-techniques-179038ea6434
3. Devika Pawar, Aakanksha Mahajan, Sachin Bhoithe. (2019) "Wine Quality Prediction Using Machine Learning Algorithms". *International Journal of Computer Applications Technology and Research* 8 (09) 385–388, ISSN: 2319-8656.
4. Ashwini Saini, Anoop Sharma. (2019) "Predicting the Unpredictable: An Application of Machine Learning Algorithms in Indian Stock Market". *Annals of Data Science.* https://doi.org/10.1007/s40745-019-00230-7
5. Yunhui Zeng, Yingxia Liu, Lubin Wu, Hanjiang Dong, Yuanbiao Zhang, Hongfei Guo, Zisheng Guo, Shuyang Wang, Yao Lan. (2019) "Evaluation and Analysis Model of Wine Quality Based on Mathematical Model". *Studies in Engineering and Technology* 6 (1), ISSN 2330-2038, E-ISSN 2330-2046.
6. Prachi Jambhulkar and Vaidehi Baporikar. (2015) "Review on Prediction of Heart Disease Using Data Mining Technique with Wireless Sensor Network". *International Journal of Computer Science and Applications* 8 (1) 55–59.
7. Samiksha H. Zaveri and Narayan Joshi. (2017) "A Comparative Study of Data Analysis Techniques in the Domain of Medicative Care for Disease Predication". *International Journal of Advanced Research in Computer Science* 8 (3) 564–566.
8. Yesim Er, Ayten Atasoy. (2016) "The Classification of White Wine and Red Wine According to Their Physicochemical Qualities". *International Journal of Intelligent Systems and Applications in Engineering* 4 (Special Issue-I) 23–26, ISSN 2147-67992147-6799.
9. https://www.kaggle.com/uciml/red-wine-quality-cortez-et-al-2009

13 Machine Vision in Industry 4.0

Applications, Challenges and Future Directions

Pramod Kumar, Dharmendra Singh and Jaiprakash Bhamu

CONTENTS

DOI: 10.1201/9781003122401-13

13.1 DEFINITION AND IMPORTANCE OF MACHINE VISION

Machine vision is an emerging technology used for the extraction of useful information automatically from digital images. In the case of manufacturing organisations, a typical machine vision process consists of manufacturing production line with a continuous and steady flow of objects. Further, it uses an automatic camera or optical system to ensure capturing of necessary details, digital photographs to analyse the captured images with a defined set of criteria. In the current scenario, the industries are using different types of vision systems for various tasks but machine vision is specifically related to industrial vision used in a manufacturing domain. Manufacturing organisations are adopting automatic machine vision instead of manual vision as this is faster, more consistent and reliable for a long period of time. Vision technology started in the 1950s but industries started the wide use of it in the 1980s.

13.1.1 MAIN FUNCTIONS OF MACHINE VISION

Machine vision system comprises the devices that are used for the collection and process of the images of physical objects that intelligently use sensors and photoelectric devices (Shan *et al.*, 2018). The core advantages of machine vision system are real-time detection, noncontact, long-term stability, remote monitoring capability, high accuracy, controllable image-taking speed and data-archiving applications, and high environmental adaptability (Li *et al.*, 2019). It has four basic functions (see Figure 13.1): measurement, counting, location and decoding (Wang *et al.*, 2019).

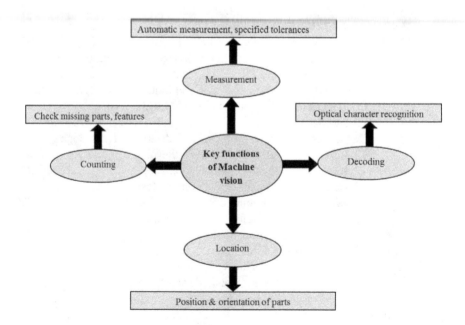

FIGURE 13.1 Main functions of machine vision.

Machine vision is widely used for automated measurements by machines that can also be checked for specified tolerances. For example, the standard gap between electrodes of the spark plugs can be better checked with close tolerances, which removes manual checking by adjusting of the gap. Machine vision is applied to find the number of parts or the number of features on a part, which allows the person to locate missing parts and also ensures proper assembly of the products. When the machine vision system is used for detecting locations, it reports the position and orientation of certain parts through a unique pattern. Decoding is the fourth function of the machine vision that refers to the decoding of two-dimensional (2D) (Data matrix, QR and Aztec) and three-dimensional (3D) symbology, i.e., linear bar codes, and stacked symbology. Decoding allows information to be recorded as historical data, which is used for immediate action.

13.1.2　Why Use Machine Vision?

Machine vision–automated technologies can bring various benefits (see Figure 13.2) to the factory floors and offers saving money and profitability increments to the manufacturers. This is accomplished in several ways by reducing defects, increasing yields, tracking parts and products and facilitating compliance with regulations. The inspection through machine vision can reduce defects and its causes, thus preventing customers receiving bad parts, which could spoil the image of the companies and can lead to costly product recalls. The vision of machines also facilitates to prevent the lines creating mislabeled product parts where the label doesn't match the content; mislabeled products develop unhappy customers and have negative impacts on a company's brand and may have safety risks. Machine vision can verify contents and ensures that products are labeled properly. It has the capability to catch defects at early stages in the manufacturing process thus reduces waste, because bad parts can be identified and eliminated before they are built into larger assemblies, which prevents scrapping of expensive materials and reworking of the parts. This increases the yield.

Machine vision helps to reduce downtime. For example, a packaging line may use automated machine vision system to detect product misfeeds that can result in jamming of the machine and thus increase down time; moreover, this vision can expel the

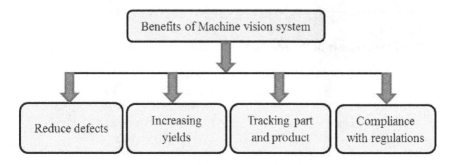

FIGURE 13.2　Benefits of the machine vision system.

product that has tolerances out of specifications. The technology of machine vision helps in tracking and tracing work-in-process as well as manufacturing throughout. Buffer stocks can be minimised and products become available more readily for just-in-time processes. Optical character reconfiguration (OCR) as well as 1D and 2D symbology can be used to track the parts and products, which helps manufacturers to avoid components shortage, reduce inventory and shorten delivery time. The automated systems of the machine vision are best suited for repetitive inspection tasks and human inspectors.

Tracking and tracing tasks are performed using dedicated bar code readers or imagers. Machine vision provides added benefits of being able to perform other tasks such as alignments or measurement in addition to reading bar codes and OCR. Complying with industry regulations is often an unavoidable cost, but if we don't comply with these regulations relating to a particular products, then the companies will not be able to take part in that market; so strict adherence to government regulations is necessary to ensure product integrity and safety.

In a broader sense, we can say the automated machine vision systems can do the things more potentially and more simply compared to human inspectors and this system never tires and works consistently on repetitive tasks with no errors.

13.1.3 KEY PARTS OF A MACHINE VISION SYSTEM

A machine vision system has five key components (see Table 13.1), including lighting, lens, communication, sensor and vision processing. Machine vision system assembly consists of the following: personal computer (PC) and an integrated smart camera along with external lighting to illuminate the parts that are built into a single device. Selection of the right configuration of the system depends on the requirements on the applications. Lighting is the main critical component as it illuminates the part to be inspected, allowing its features to stand out so that the camera can clearly see them. A lens captures the image and presents it to sensor in the form of light. The sensor is a machine vision camera and converts this light into a digital image that is sent to a processor for analysis. Vision processing consists of algorithms that review the image and extract the required information the system will run whatever measurements or other processes are. The final data is

TABLE 13.1

Key Components of the Machine Vision System

S. No.	Name of Component	Function
1.	Lighting	Illuminates the part to be inspected
2.	Lens	Captures the image and presents it to sensor
3.	Communication	Data communication in a useful manner
4.	Sensor	Converts light into a digital image
5.	Vision processing	Algorithms that review the image and extract the required information

instructed to run and is communicated in a useful manner. Placement and orientation of the products should be consistent and repeatable to achieve the best possible results. The position and type of light should be carefully selected to maximise contrast of the features. The machine vision system uses vision processing that is the core of the system to extract the information that is required for an application process. Camera vision processing is performed by software with a wide variety of available interfaces and tools.

Some common vision software tools include locating, counting, measuring and decoding tools. These tools are used in a sequence to achieve a desired result. The purpose of machine vision system is to extract useful information from digital images; the communication of that data is critical and this is done by either a discrete input-output (I/O) signal or data sent over a serial connection to a device that's logging information or using in many smart cameras to have these connections built into them. Discrete I/O points can be connected to either a programmable logical controller (PLC) which will use that information to control a work cell or an indicator such as stack light or directly to a solenoid that might be used to trigger reject mechanism data communication by a serial connection. Some systems employ a higher level industrial protocol, which might be connected to a device like a human machine interface (HMI) screen to provide visual confirmation that the system is running.

13.1.4 STEPS IN MACHINE VISION

The first step in machine vision is image acquisition consisting of a camera and illumination. The properties of illumination are used to increase the individual properties of features that are to be observed. Three-dimensional scenes are observed through a lens and light reflected by this 3D scene is actually passing by or passing through the lens system and these light rays strike the surface of a pixelised sensor. Each pixel contains information about the intensity of light and color information of light that was reflected back from this 3D scene; then, we have 2D representation, which is a projection of a 3D object.

Preprocessing is the computational step that we perform after acquiring an image from a camera sensor. Preprocessing is needed because of uneven illumination of a scene and this background shading is to be removed. Images can also be polluted by noise and these noise statistics further disturb the computational step of later processing of the machine vision system. We need to apply a kind of operation to clean up the noise of the image.

Segmentation helps to find interesting image objects and separate them from the background. This is done by using a gray scale image or a color image that we are retrieving from the preprocessing step then looking into statistics of each pixel and making this decision based on a gray scale value or a color value, if this pixel belongs to any kernel of an image object or to background, and then doing the segmentation. One important conclusion is to reduce the amount of data because data representation from a segmented image actually depicts values of being an object or a background, which means that we have the binary pixel either its one or its zero when compared with a gray scale or a color image, then the each pixel is represented by an integral and the depth of integral is done by the size of it.

Component labeling is the computational step where we take a binary segmented image as input and this image contains a number of objects of interest; these objects are considered media and each object is assigned a unique code by scanning procedure; this unique code is visualised by its own unique color code.

After component labeling, there is need to extract certain properties of the image that will be a description of each component, which is equal to the area of the image component that can be sub-pixel position. At the end of the process, a statistical classifier is used for further action.

13.2 CONCEPT AND DEFINITION OF INDUSTRY 4.0

Industry 4.0 also has emerged as the fourth industrial revolution in global manufacturing, which is based on the Internet of Things (IoT) concept that has integrated the current production systems vertically as well as horizontally (Thoben *et al.*, 2017). Traditional factories are now transforming to smart factories and these factories now have capabilities to address dynamic and variable demands of the ever-changing preferences of customers adopting automation with intelligent manufacturing. To map with the current changing scenario and global competitiveness, the manufacturing organisations of different countries are striving to develop research and technological schemes. Germany is enacting its Industry 4.0 to cope with European policy while the United States has started smart manufacturing. Japan and Korea have also commenced their initiatives of smart manufacturing.

Industry 4.0 has objectives to make manufacturing intelligent based on cyber-physical systems (CPS). Machines, services, products and all entities are equipped with advanced technologies such as smart sensors, radio-frequency identifications (RFIDs), microprocessors and other embedded systems that enable them to collect, evaluate, analyse data and develop capacities to connect and communicate with other devices and environment.

13.2.1 THE EVOLUTION OF INDUSTRY 4.0

The evolution of industrial revolutions (see Figure 13.3): the first industrial revolution was started at the end of the 18th century, which was based on water and steam power generation. During this, mechanical production equipment were driven by the power extracted from steam and water. It was beginning of the 20th century, when the second industrial revolution was coined with the division of labor concept. This revolution was based on mass production with the use of electric power. Material handling through a conveyor belt was also utilised during the second revolution.

In this series, the third industrial revolution was coined in the beginning of the 21st century, which was based on robotics and electronics automation. Computers, numerical controller (NC) and PLC were the core of the third industrial revolution. The fourth industrial revolution was coined in Germany in 2013, but nowadays, other European countries are also adopting this new manufacturing paradigm (Vuksanovi *et al.*, 2017). This new shift of manufacturing is dependent on the digitalisation of the entities. CPS and smart automation are the backbones of the emerging technologies of the fourth industrial revolution. Companies are moving toward more

Mechanical automation	Industrialization	Microelectronics Robotics	Digital processes & ICT
Engineering Science	Conveyor belt Mobility	Computer, NC, PLC	Cyber-Physical System
Water & steam Power	Mass production	Electronic automation	Smart automation
Industry 1.0	Industry 2.0	Industry 3.0	Industry 4.0
End of 18th century	Beginning of 19th century	Beginning of 21th century	2013

FIGURE 13.3 The revolution of Industry 4.0.

flexibility and productivity to achieve better quality of products and services with the use of disruptive technologies of Industry 4.0.

13.2.2 DESIGN PRINCIPLES OF INDUSTRY 4.0

Design principles (see Figure 13.4) are the foundation of Industry 4.0 that supports factories to identify its pilots and to develop appropriate solutions (Hermann *et al.*, 2016). Following are the design principles of the fourth industrial revolution.

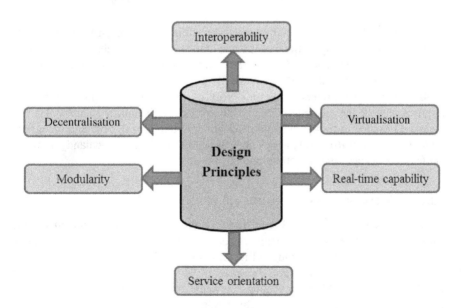

FIGURE 13.4 Design principles of Industry 4.0.

Interoperability: This refers to the capability of various machines, people and products to interact for data sharing and coordination of information. This coordination is maintained through IoT and CPS. Interoperability is classified into four categories, i.e., structural, semantic, system and syntactic. Interoperability is the core principle used to establish connectivity between smart products, smart factories, people and services (Lu, 2017). The connection between CPS is maintained by structural interoperability standards and concepts used to exchange data in a uniform manner. An IoT platform helps in linking information and operational technologies. To establish an intelligent way of production, there is always need to integrate data between men-machine and machine-machine in an efficient manner; therefore, to achieve better efficiency, companies are shifting on digital operations.

Decentralisation: In the current scenario of manufacturing, centralisation is no longer necessary, so all machines should work independently and make decentralised decisions. Customer's demands are personalised, so it is difficult to control systems centrally. Different components inside a smart factory can work under self-control by using decentralised concept. Embedded computer system makes CPS with intelligent decision-making (Hompel and Otto, 2014).

Virtualisation: This principle belongs to create a virtual copy of a physical entity. The entire supply chain of a company is now replaced by virtual models that are simulation-based models (Moreno *et al.*, 2017). The virtual model of the physical objects can work as preventive signification and indicates the defects before they occur. This will reduce the maintenance and service costs of the objects.

Real-time capability: Real-time data collection is necessary for quick response and this is possible through Big Data and IoT. In an Industry 4.0 environment, data exchange between various entities such as factory, smart warehouse, business partners and products is supported by the Internet of Everything (IoE) (Lee *et al.*, 2015; Zhang *et al.*, 2017).

Service orientation: Companies have interlinked their service facilities with manufacturing. The applications of IoT and cloud computing enable customer communication with services provided by the companies. RFID tags are used to build a service-oriented structure of the industries.

Modularity: This refers to the flexibility of changing the modules as per the requirements of the customer. This is the core phenomenon of Industry 4.0 that makes it possible to transform rigid production system into flexible and reconfigurable (Gilchrist, 2016). Modularity copes up with ever-changing demands of different features, capabilities, functions and attributes.

13.2.3 TECHNOLOGIES OF INDUSTRY 4.0

Industry 4.0 technologies (see Figure 13.5) are the pillars of current manufacturing shift, which help to achieve higher operational efficiency as well as flexibility in operations. Following are the technologies that make building blocks of the manufacturing.

Internet of Things (IoT) and Internet of Services (IoS): The Internet is the backbone for establishing connections between different things and services. All the devices are linked to a cloud to achieve solutions for analytics and computations (Erboz, 2017). Agility and flexibility in manufacturing operations are achieved via

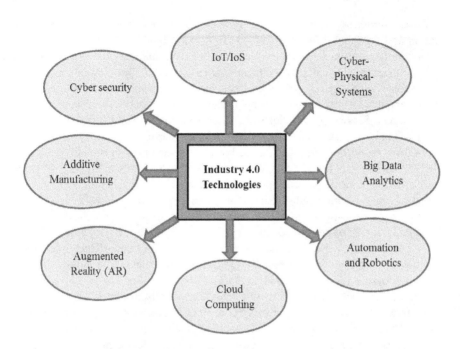

FIGURE 13.5 Technologies of Industry 4.0.

IoT capabilities of firms by integrating the facilities through information and communication devices. The modularity of the production processes is done by interconnecting smart sensors, smart machines, physical entities and human factors. IoT infrastructure enables the accessibility of smart, reliable and autonomous products and services across the value chain (Schumacher *et al.*, 2016).

Cyber-physical systems: This is one of the core technologies to assimilate physical objects, machines and humans via the Internet using computer-based algorithms (Lee *et al.*, 2015; Gilchrist, 2016). Performance of the machines in the factories that can be enhanced by digital twining of physical and virtual models and monitoring and controlling is predicted. CPS-enabled supply chains have greater advantages over traditional ways and lead to potential reforms.

Big Data analytics: This is the technology that helps in collecting, interpreting and analysing the data extracted from smart devices so that faults can be predicted before occurrence of the faults. Big Data analytics develops capabilities to take better decisions based on the information gathered from fact and data. A large variety and volume of data are managed for identifying useful insights to take smart decisions. Big Data analysis discovers, captures and analyses extracted data to achieve optimal results.

Automation and robotics: Robots and automated machines are essential requirements of current manufacturing organisations to address the trends of flexible manufacturing. Therefore, adaptive and flexible robots are working in assembly and product development. These robots can work in all environments of companies and complicated tasks can be performed more precisely and safely (Salkin *et al.*, 2018). Intelligent robots are used for packaging, welding and other sensitive tasks.

Cloud computing: Companies are using cloud-based technologies with an information and communication technology (ICT) paradigm to facilitate management and administration tasks more easily. Manufacturing and services are interlinked with the use-available cloud-based technologies (Ooi *et al.*, 2018). Machines, equipment and devices on the shop floor are connected to the same cloud for sharing information about the status and monitoring. Different cloud-based service platforms, i.e., Platform as a Service (PaaS), Software as a Service (SaaS) and Infrastructure as a Service (IaaS), are used by industries for providing services.

Additive manufacturing: Customised product demands of consumers can be fulfilled by adopting an additive manufacturing approach. Companies using 3D printing facilities are moving toward producing personalised products on time with the lowest cost and a lightweight design (Rubmann *et al.*, 2015). Various methods of additive manufacturing, such as laser sintering, fused deposition, are the core behind the manufacturing of customised products. In these methods, layer-by-layer deposition is performed to achieve the desired shape and design of the product in less time. Geometry of CAD models are read by additive machines and these models are converted to physical objects (Esmaeilian *et al.*, 2016).

Augmented reality (AR): This represents the visualised models that show real-time status of working of machines. The process behavior of machines is also reflected by these visualised models (Zheng *et al.*, 2018). AR is a visualisation technique that is used for the real-time monitoring of manufacturing processes as well as fault detection of products. It offers an establishment of connection between human-machine for controlling maintenance tasks (Elia *et al.*, 2016).

Cybersecurity: Everything is connected via the Internet to maintain communication between different devices so there always exist risks of data leakage and cyber-attacks within or outside the boundaries of factories. Therefore, organisations need to adopt standard, safe and reliable protocols to establish communication in Industry 4.0 environment (Rubmann *et al.*, 2015).

13.3 MACHINE VISION IN THE AGE OF INDUSTRY 4.0

Industrial components are continuously becoming smaller in size, and products get to market with less waste, faster and on-time. Even most manual processes are turning toward automation. Innovations in machine vision, deep learning and robotics are revolutionising production lines and supply chains. The machines now produce a vast amount of Big Data accessible via the cloud. As machines and the information they produce become linked, new CPS will spark a modern-day industrial revolution. This is called Industry 4.0. In Industry 4.0, smart machines learn from their environments and take corrective actions to optimise production. These machines work independently of a central controller, collaborating and communicating with other devices. Industry 4.0 factories produce new organisational intelligence that spreads across facilities and is accessible anywhere.

Machine vision is the heart of this manufacturing paradigm; many companies are offering Industry 4.0–ready solutions to help manufacturers innovate for future and transmit data online to enable it to communicate with other devices to transfer manufacturing data. Different PLCs can be linked with factory's networked devices for data collection across plans and global networks and insights into operations and performance improvements. Sensors guide and communicate with collaborative robots that work safely alongside humans. Real-time monitoring capabilities anticipate and accommodate changes and adjust based on incoming data.

13.3.1 Machine Vision Accelerating Industry 4.0 Paradigm

Machine vision is the emerging trend in technology that combines cameras, sensors and software that act as 'eyes' for new manufacturing revolution. Industry 4.0 is driving manufacturing, services systems using automated, interconnected artificial intelligence (AI) and CPS (Coffey, 2018), also known as Industrial IoT (IIoT). There is a vast use of machine vision in the age of Industry 4.0. Therefore, the manufacturers are potentially using machine vision due to technological advancements and continuous demand of customised products at lower cost (Zancul *et al.*, 2020).

13.3.1.1 Applications of Machine Vision in the Context of Industry 4.0

In the current manufacturing era, to foster productivity, flexibility and on-time delivery, the industries are transforming toward advanced manufacturing so these are adopting machine vision technologies in a wide range of applications (see Figure 13.6).

Machine vision technologies have been used for the inspection of quality of components and sorting of the desired products (Zancul et al., 2020). Companies are

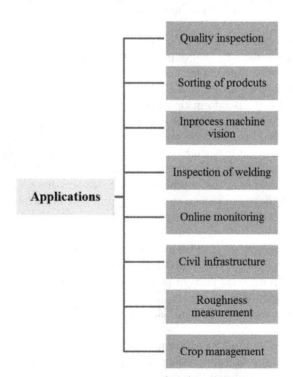

FIGURE 13.6   Applications of machine vision systems.

developing and using machine vision–based systems to design assembly stations to improve worker ergonomy. Machine vision technologies also assist in lean manufacturing flow with an integration of MES and ERP software.

Machine vision technologies based on CPS are also used for control and monitoring of the tool wear in machining processes in manufacturing industries (Lins *et al.*, 2020). The integration of in-process machine vision in production systems can enhance feasibility and effectiveness of the machining processes being carried out.

Rapid developments in machine vision and Industry 4.0 technologies make great contributions to the manufacturing sector by providing automation and flexibility in production, inspection, planning and control. The manufacturing sector is achieving new heights with the innovation of new emerging technologies, including machine and deep learning, image-processing and data transmission capabilities with standard protocols, complementary metal-oxide semiconductor (CMOS) sensors, embedded vision and interfaces. For example,

  a. The chemical composition of the materials can be determined using the hyper spectral image technique
  b. Stress pattern in the materials can be accessed using the polarisation image method
  c. In production lines, trends in critical measurements can be checked and analysed by retrofitting machine vision systems

Machine vision applications and common challenges are mentioned in Table 13.2. Machine vision is used in controlling different parameters of welding. It is also applied to determine transient temperature distribution in the submerged arc welding (SAW) technique (Sarkar *et al.*, 2020). This system captures real-time infrared images and extracts information data regarding a transient shape of isotherm of plates. This contactless technique is also used to determine welding characteristics such as cooling rate patterns, maximum temperature and isotherm generation. Infrared-based cameras in machine vision systems offer a fast reliable determination of temperature without making physical contact in the hot environment of welding processes. Lei *et al.* (2019) machine vision-based crosslines laser to control and measure of weld height in sheet metal welding process when a tube is to be welded with tube-sheet. Adaptive correction and control is possible with this system to avoid thermal distortion.

Machine vision systems have been applied to recognise different parts of products by using color lighting vision technologies. Machine vision approach is applied to identify or detect different parts of pen, i.e., cap, tube and bottom by using color-based various lighting (Bagheri *et al.*, 2020). Machine vision system is developed to guide and direct the robots parts to take accurate decisions for selecting the color-based parts.

Advancements in machine vision technologies explore the doors in agriculture area for quality inspection and grading. Many plant diseases can be recognised using these disrupting technologies. For example, farmers are facing challenges in detection as well as classification of diseases in papaya plants. Papaya disease is recognised

TABLE 13.2
Machine Vision Applications and Common Challenges

Machine Vision Principles/Techniques	Applications	Common Challenges
Bar code scanner	1. In inventory control, products can be monitored during storage and distribution in retail industries 2. Customer service applications 3. Automatic check out systems	Lack of system engineering skills
Visual data detection using sensors	Predictive maintenance systems: For example, sensors for heat and vibrations detection	Performance of machine vision-based systems is affected by environmental factors, i.e., lighting conditions and illumination changes
Rapid processing capability	High-quality production control, quick identification of defective parts and reduced human errors	Requirements of high-resolution sensors and lens for specific applications, i.e., drone has to capture a wider area
Ultrasonic techniques	Healthcare industries: skin cancer detection, surgical navigation, etc.	Competing requirements
Automated data capture and automated inspection	Logistics: To control and verify the products using QR codes, RFIDs, etc.	Need embedded devices that are small in size and weight, highly specified systems for image processing
Automotive	Dimensional gauging-check geometric positions as per specifications	Challenges such as the amount of data, data bandwidth, demand for high performance—low power and heterogeneous computing

using online agro-medical expert systems based on machine vision (Habib *et al.*, 2018), the images of which can be captured through cameras, mobiles, which is the process to identify the disease to address the problem. This machine vision–based agro-medical expert system can detect the features from images to access the contamination of disease in papaya. Machine vision–based hardware is developed along with its LabVIEW software for estimating the weight of potatoes and classifying its categories according to weights as per Philippines National Standards (Quilloy et al., 2020).

Machine vision system is also used for the roughness measurement of soil and for improvement in soil conditions to handle inhomogeneous properties of soil. A stereo camera–based machine vision system is developed to detect soil roughness using regulating mechanism for the speed of power harrow (Riegler-nurscher *et al.*, 2020). Machine vision technologies have also shown their contributions in civil infrastructure for structural health monitoring (Tang *et al.*, 2020). Condition of civil infrastructures status, i.e., dams, highways, bridges and reservoirs, are evaluated by using smart sensors (Zhu *et al.*, 2011). Disrupting technology such as

AI can be adopted for intelligent city planning, 3D printing and building information modeling.

Manufacturers can use machine vision to monitor wear of the tools to avoid damage (Saglam, 2011). Wang *et al.* (2020) developed a machine vision–based system for the monitoring of cutting wheel degradation for LCD screen–cutting process. This process is performed by estimating the height of the blades of the cutting wheel by extracting images. Monitoring of tool wear is necessary to predict the useful life of the inserts.

Machine vision has been applied to detect various quality parameters, i.e., size, shape, color, texture of fruits and vegetables. Srivastava et al. (2020) introduced sensor-based hand-held machine vision system to check different quality parameters such as chlorophyll, weight, volume and sugar content of citrus cultivars. Effective crop management is possible by using machine vision system. Kim *et al.* (2019) developed field monitoring system for automatic crop monitoring that can extract the information of disease symptom of onion downy mildew. This automatic field-monitoring system contains cameras, wireless transceivers, image logging modules and a motor system. Fan *et al.* (2020) designed online cost-effective computer vision system based on deep learning techniques to identify defective apples on fruit sorting machine.

Frustaci et al. (2020) designed and developed embedded machine-based system for inline geometric inspection of catalytic converter assembly process for flexible, low cost and precise inspection. Afterward, defects in geometry due to rotational shifts of interfaces about their mean positions can also be detected. Real-time counting of manually assembled components is carried out by using versatile algorithm of machine vision (Pierleoni *et al.*, 2020). This system has capability to deal with human interactions and counting of assembled pieces by using real-time video inputs. It consists of inter-frame analysis, image preprocessing, binarisation, morphological operations and blob detection. One custom detector based on machine learning is considered a reference for comparison.

Machine vision can also be used for detecting the leakage of harmful gases that can disturb the climate. Wang *et al.* (2020) developed and used optical gas imaging (OGI) technique based on computer vision with an infrared camera to detect the leakage of natural methane gas emissions. OGI method is a widely used process for leakage detection but it is labor-intensive, but if this is attached to machine vision system with convolutional neural networks (CNN), then its efficiency increases.

Coal column flotation circuit monitoring and analysis is possible using machine vision methods (Massinaei *et al.*, 2018). This system is used to diagnose the process conditions and predict the performance of it in different operating conditions. Surface defects and irregularities can also be automatically detected using machine vision–based systems. Scratches and dents occur during manufacturing and transporting on the automobile surfaces, need to be removed before reaching customers so the auto-inspection system is designed, developed and implemented to detect the surface defects preset on handles, edges and style lines of automobiles (Zhou *et al.*, 2019). This system works on image acquisition and processing principles.

13.4 CHALLENGES IN THE PATH OF MACHINE VISION AND INDUSTRY 4.0

13.4.1 CHALLENGES IN THE IMPLEMENTATION OF MACHINE VISION TECHNOLOGIES

Companies are NOW focusing on intelligent autonomy and high-performance operating for applications such as computer vision that is very challenging application that needs high-level performance and reliability. An image-processing computer vision is applied everywhere or all over the world nowadays, i.e., safety critical systems, medical, logistic and traffic control, avionics and space, drones and autopilots, navigation and car safety, robotics and exoskeletons, power line monitors and virtual reality. Therefore, we are facing challenges such as an increasing number of sensors, amount of data, data bandwidth, demand for high performance—low power and heterogeneous computing, embedded devices in small in size and weight, highly specified systems for image processing. These challenges include the demand for intensive image process, embedded system constraints (low power, size limits and weights) and time to market or cost sensitiveness of devices.

Practical issues also exist while using machine vision systems for different applications. For example, machine vision–smart cameras are used to locate, guide, check and measure the objects in different robot applications to perform various tasks in industry, i.e., pick-place, assembly, packaging and palletising, quality inspection, screw driving, lab analysis and testing, labeling, polishing, injection molding. The following practical issues have been seen while using these smart cameras in robot applications:

a. While using machine vision with higher value operation in robot applications, there is a requirement of higher speed and accuracy.
b. Issues exist related to figure out the proper space, location to fit and mount the smart cameras.
c. High flex cables are required to complete the millions of cycles.
d. Need simple programming to complete the desired task.
e. Requirement of system calibration for 2D and 3D motions.
f. Common protocols are always needed to access the data from machine vision system by the robots.

The above-mentioned practical issues are limited to robot applications. There may be other issues while using machine vision system in different domains such as requirement of high-resolution sensors, proper wavelength for light emitted, lens design. For example, in modern drone applications there is need to capture a wider area within limited flight duration so high-resolution sensors are required to solve this purpose, which leaves challenges before lens designers. Therefore, practitioners should understand those issues before installing machine vision system in their areas of applications.

13.4.2 CHALLENGES IN THE IMPLEMENTATION OF INDUSTRY 4.0 TECHNOLOGIES

Industry 4.0 technologies help in customised, effective, flexible and automated production economically by using smart sensors, machines and other advanced ICT.

Manufacturers need to adopt specific strategies for proper implementation of these disrupting technologies in their new business models, but this is a big challenge for companies. Therefore, the practitioners should expect to design and develop the infrastructure having capabilities to address these barriers in the implementation of digital technologies. Industry 4.0 offers many benefits, but it is the barriers that create obstacles before the practitioners in successful adoption of these emerging technologies. Therefore, it is important to understand these barriers by exploring ways to remove them (Suresh *et al.*, 2018).

Researchers in this domain have identified core challenges, i.e., high investment, lack of adequate knowledge, cyber threats, lack of infrastructures. Driving and dependency patterns are also known for better understanding the interrelationship between these barriers. The following are the challenges in adoption of Industry 4.0 in current manufacturing organisations.

13.4.2.1   Poor Value-Chain Integration

Industry 4.0 basically requires the seamless connections across entire value chains. There should not be present any miscommunication between departments in factories as well as external partners of value chains. Poor coordination is the core challenge for companies in the implementation of advanced technologies in their production and services. To achieve efficient cyber-physical connectivity, companies are facing problems in interoperability in an Industry 4.0 environment. To maintain smooth coordination and connection, there is always necessity to establish horizontal, vertical and end-to-end integration through IoT.

13.4.2.2   Cybersecurity Threats

All the tasks in Industry 4.0 environment are carried out using connectivity of the all physical entities via the Internet. Therefore, there are always chances of leakage of information. Organisations planning to adopt advanced technologies must take effective measures to prevent data leakage problems so that the risk of data security can be eliminated (Aggarwal *et al.*, 2019; Moeuf *et al.*, 2019).

13.4.2.3   Confusion about Economics Results

Industry 4.0 is still in a transition phase so companies doubt its economic benefits and there is no assessment mechanism to understand the practical way of receiving advantages of invested capital for implementing these disrupting technologies. Therefore, organisations are facing difficulties in predicting the positive benefits of implementation (Kamble *et al.*, 2018; Moeuf *et al.*, 2019).

13.4.2.4   Lack of Required Skills

Industry 4.0 is based on digital technologies and intelligent principles so there is need of adequate skills in workforce. Qualified workers having knowledge of digital and advanced technologies will be in demand while shifting toward this new phase of manufacturing. Lack of adequacy in digital techniques is one of the key challenges in adopting Industry 4.0 technologies (Raj *et al.*, 2019).

13.4.2.5 Need of High Investments

For implementing Industry 4.0 technologies, industries are facing funding problems. Most of the SMEs don't have enough capital for the investment required to implement these disrupting technologies (Kamble *et al.*, 2018). There is always a need for huge capital to shift toward advanced technologies.

13.4.2.6 Lack of Infrastructure

Industries are facing a shortage of adequate digital and IT infrastructure facilities. Industry 4.0 requires high speed Internet, computer labs for a proper adoption of the advanced techniques of production. Lack of Internet coverage, IT infrastructure and other facilities, i.e., electricity, water, can hinder the progress of Industry 4.0 (Anand and Nagendra, 2019).

13.4.2.7 Loss of Jobs

In an Industry 4.0 environment, automation with intelligent robotics is used for manufacturing and services, so practitioners are anticipating loss of jobs. Advancement in technologies will disrupt jobs, which will lead to a big challenge in labor market (Khan and Turowski, 2016).

13.4.2.8 Challenges in Data Handling

Industries planning to transform their production lines with respect to disrupting technologies need to arrange facilities to manage a vast amount of data. Sensors and machine products generate a huge amount of data that has to be handled in a right way to extract valuable information. Modern data analysis, data mining techniques are the basic requirements of Industry 4.0.

13.4.2.9 Need Secure Standards and Norms

Connection and coordination among industries is a basic need of Industry 4.0, so common standards and regulations should be adopted. There is a lack of uniform protocols and standards that are the challenges before an organisation to integrate its value chains (Raj *et al.*, 2019). Integrating value-chain networks with different standards can hinder the connectivity of industries. So companies should decide reference standards for a proper integration of facilities.

13.4.2.10 Resistance to Change

Workers always resist shifting to advanced technologies as they are worried about losing their jobs. Even small organisations are feeling resistance to shift toward disrupting technologies-based production. Employees are not ready to adopt these technologies and are showing unwillingness.

13.5 CONCLUSION AND FUTURE DIRECTIONS

In this chapter, the fundamental concepts of machine vision and Industry 4.0 are explained with their applications and challenges. Machine vision technologies are the eyes of the Industry 4.0 so organisations are trying to adopt both of the technologies to map with current manufacturing shifting. Machine vision steps are also elaborated in this chapter and the applications of machine vision in different industry sectors are described. Machine vision and Industry 4.0 exist together and support each other. Machine vision methods are used for a wide variety of applications such as inspection of parts, monitoring of welding parameters, disease recognition, quality checking and reducing chances of failures. Machine vision systems use sophisticated technologies and powerful hardware to access information about characteristics of products and processes.

Machine vision technologies have substantially emerged in last 15 years and have been used for automation in manufacturing industries. Advancements in these systems have explored new ways in electronics, automotive, medical, semiconductor, pharmaceuticals and packaging industries. In the context of Industry 4.0, machine vision systems have been used to measure different dimensions, guiding robots, machines, identifying parts and inspecting assembly operations and inspecting defective parts in production lines with noncontact means.

However, continuous advancements in performance, cost, algorithmic robustness and ease of use will be the base of adopting machine vision technologies in manufacturing automation. In future, more advances at shop floors in this series are expected that will characterise the status of industrial machine vision in next few years. The future vision of Industry 4.0 can be described by the following characteristics: fast speed, intuition, intelligence and ease of use, to be used for component to component or process variations. While the application of machine vision in Industry 4.0 environment might not have reached the mark but recent inventions have helped us to accelerate the use of machine vision systems in current industries. Despite that, academic institutions and governments need to start initiatives to make the people aware of the economic benefits of using machine vision with Industry 4.0 and should provide the right platform for the same. Also, training program and curriculum must be designed to develop adequate skills for these emerging technologies.

REFERENCES

Aggarwal, A., Gupta, S. and Ojha, M. K. 2019. Evaluation of key challenges to Industry 4.0 in Indian context: A DEMATEL approach. In: Shanker K., Shankar R., Sindhwani R. (eds) Advances in Industrial and Production Engineering. Lecture Notes in Mechanical Engineering. Springer, Singapore. 387–96.

Anand, P. and Nagendra, A. 2019. Industry 4.0: India's defence industry needs smart manufacturing. International Journal of Innovative Technology and Exploring Engineering. 8(11S):476–85.

Bagheri, N. M., van de Venn, H. W. and Mosaddegh, P. 2020. Development of machine vision system for pen parts identification under various illumination conditions in an Industry 4.0 environment. 1:1–20. doi: 10.20944/preprints202004.0387.v1

Coffey, V. C. 2018. Machine vision: The eyes of Industry 4.0. Optics and Photonics News. 29(7): 42–9.

Elia, V., Gnoni, M. G. and Lanzilotto, A. 2016. Evaluating the application of augmented reality devices in manufacturing from a process point of view: An AHP based model. Expert Systems with Applications. 63(1): 187–97.

Erboz, G. 2017. How to define Industry 4.0: The main pillars of Industry 4.0. In: Proceeding of 7th International Conference on Management (ICoM 2017) at Nitra, Slovakia: 1–9.

Esmaeilian, B., Behdad, S. and Wang, B. 2016. The evolution and future of manufacturing: A review. Journal of Manufacturing Systems. 39(1): 79–100.

Fan, S., Li, J., Zhang, Y., Tian, X., Wang, Q. and He, X. 2020. On line detection of defective apples using computer vision system combined with deep learning methods. Journal of Food Engineering. doi: 10.1016/j.jfoodeng.2020.110102.

Frustaci, F., Perri, S., Cocorullo, G. and Corsonello, P. 2020. An embedded machine vision system for an in-line quality check of assembly processes. Procedia Manufacturing. 42: 211–8.

Gilchrist, A. 2016. Industry 4.0: The Industrial Internet of Things. Springer, Heidelberg.

Habib, T., Majumder, A., Jakaria, A. Z. M., Akter, M. and Uddin, S. 2018. Machine vision based papaya disease recognition. Journal of King Saud University Computer and Information Sciences. 0–9. doi: 10.1016/j.jksuci.2018.06.006.

Hermann, M., Pentek, T. and Otto, B. 2016. Design principles for industrie 4.0 scenarios. In: Proceeding of the 49th Hawaii International Conference on System Sciences (HICSS), IEEE, Koloa, HI, New York City, NY: 928–37.

Hompel, T. M. and Otto, B. 2014. Technology for versatile logistics. Industry 4.0. In: Proceeding of 23rd German Material Flow Congress.

Kamble, S. S., Gunasekaran, A. and Sharma, R. 2018. Analysis of the driving and dependence power of barriers to adopt Industry 4.0 in Indian manufacturing industry. Computers in Industry. 101: 107–19. https://doi.org/10.1016/j.compind.2018.06.004

Khan, A. and Turowski, K. 2016. A survey of current challenges in manufacturing industry and preparation for Industry 4.0. In: A. Abraham et al. (eds.), Proceedings of the First International Scientific Conference "Intelligent Information Technologies for Industry" (IITI'16). Advances in Intelligent Systems and Computing. 450:. 15–26. https://doi.org/10.1007/978-3-319-33609-1_2.

Kim, W., Lee, D. and Kim, Y. 2019. Machine vision-based automatic disease symptom detection of onion downy mildew. Computers and Electronics in Agriculture. 168: 1–10.

Lee, J., Bagheri, B. and Kao, H. A. 2015. A cyber-physical systems architecture for Industry 4.0-based manufacturing systems. Manufacturing Letters. 3(1): 18–23.

Lei, T., Wang, W., Rong, Y., Xiong, P. and Huang, Y. 2019. Cross-lines laser aided machine vision in tube-to-tubesheet welding for welding height control. Optics and Laser Technology. https://doi.org/10.1016/j.optlastec.2019.105796.

Li, S., Xuefeng, Z. and Guangyi, Z. 2019. Automatic pixel-level multiple damage detection of concrete structure using fully convolutional network. Computer-Aided Civil and Infrastructure Engineering. 34(7): 616–34.

Lins, R. G., de Araujo, P. R. M. and Corazzim, M. 2020. In-process machine vision monitoring of tool wear for cyber-physical production systems. Robotics and Computer Integrated Manufacturing. 101859. doi: 10.1016/j.rcim.2019.101859.

Lu, Y. 2017. Industry 4.0: A survey on technologies, applications and open research issues. Journal of Industrial Information Integration. 6(1): 1–10.

Massinaei, M., Jahedsaravani, A., Taheri, E. and Khalilpour, J. 2018. Machine vision based monitoring and analysis of a coal column flotation circuit. Power Technology. 343: 330–41.

Moeuf, A., Lamouri, S., Pellerin, R., Tamayo-Giraldo, S., Tobon-Valencia, E. and Eburdy, R. 2019. Identification of critical success factors, risks and opportunities of Industry 4.0 in SMEs. International Journal of Production Research. 58(5): 1–17.

Moreno, A., Velez, G., Ardanza, A., Barandiaran, I., de Infante, Á. R. and Chopitea, R. 2017. Virtualisation process of a sheet metal punching machine within the Industry 4.0 vision. International Journal on Interactive Design and Manufacturing. 11(2): 365–73.

Ooi, K.-B., Lee, V.-H., Tan, G. W.-H., Hew, T. S. and Hew, J. J. 2018. Cloud computing in manufacturing: The next industrial revolution in Malaysia. Expert Systems with Applications. 93(1): 376–94.

Pierleoni, P., Belli, A. and Palma, L. 2020. A versatile machine vision algorithm for real-time counting manually assembled pieces. Journal of Imaging. 6(6). doi: 10.3390/jimaging6060048.

Quilloy, E. P., Sanchez, P. R. P., Manuel, L. J. R. and Renovalles, E. M. 2020. Development of machine vision system for size classification of potatoes (*Solanum tuberosum* L.). Philippine Journal of Agricultural and Biosystems Engineering. 15(2): 23–30.

Raj, A., Dwivedi, G., Sharma, A., Lopes de Sousa Jabbour, A. B. and Rajak, S. 2019. Barriers to the adoption of Industry 4.0 technologies in the manufacturing sector: An inter-country comparative perspective, International Journal of Production Economics. 224(C):1–43.

Riegler-nurscher, P., Moitzi, G., Prankl, J., Huber, J., and Karnel, J. 2020. Machine vision for soil roughness measurement and control of tillage machines during seedbed preparation. Soil and Tillage Research. 104351. https://doi.org/10.1016/j.still.2019.104351.

Rubmann, M., Lorenz, M., Gerbert, P. and Waldner, M. 2015. Industry 4.0: The future of productivity and growth in manufacturing industries. https://image-src.bcg.com/Images/Industry_40_Future_of_Productivity_April_2015_tcm9-61694.pdf (BCG group)-April 2015.

Saglam, H. 2011. Tool wear monitoring in band sawing using neural networks and Taguchi's design of experiments. The International Journal of Advanced Manufacturing Technology. 55: 969–82.

Salkin, C., Oner, M., Ustundag, A. and Cevikcan, E. 2018. A conceptual framework for Industry 4.0. In:Pham D.T. (eds.)Industry 4.0: Managing the Digital Transformation. Advanced Manufacturing. Springer, Cham. 3–23.

Sarkar, S. S., Das, A., Paul, S., Ghosh, A., Mali, K., Sarkar, R. and Kumar, A. 2020. Infrared imaging based machine vision system to determine. Infrared Physics and Technology. 103410: 1–32. doi: 10.1016/j.infrared.2020.103410.

Schumacher, A., Erol, S. and Sihna, W. 2016. A maturity model for assessing Industry 4.0 readiness and maturity of manufacturing enterprises, Changeable, Agile, Reconfigurable & Virtual Production, Procedia CIRP Paper. 52: 161–6. Available at https://publik.tuwien.ac.at/files/publik_255446.pdf

Shan, B., Yunfeng, P. and Xiaoyang, H. 2018. Detection of slip for CFRP-concrete interface using stereovision method corrected by epipolar constraint. Structural Control and Health Monitoring. 25(10): e2212.

Srivastava, S., Vani, B. and Sadistap, S. 2020. Machine vision based handheld embedded system to extract quality parameters of citrus cultivars. Journal of Food Measurement and Characterization. https://doi.org/10.1007/s11694-020-00520-2. Springer US, (0123456789).

Suresh, N., Hemamala, K. and Ashok, N. 2018. Challenges in implementing industry revolution 4.0 in Indian manufacturing SMES: Insights from five case studies. International Journal of Engineering & Technology. 7(2.4): 136.

Tang, Y., Lin, Y., Huang, X., Yao, M., Huang, Z. and Zou, X. 2020. Grand challenges of machine-vision technology in civil structural health monitoring, Artificial Intelligence Evolution, 1(1): 8–16.

Thoben, K.-D., Wiesner, S. and Wuest, T. 2017. Industry 4.0 and smart manufacturing – A review of research issues and application examples. International Journal of Automation Technology. 11(1): 1–12.

Vuksanovi, D., Ugarak, J. and Korcok, D. 2017. Industry 4.0 – The future concepts and new visions of factory of the future development. Advanced Engineering Systems, International Scientific Conference on ICT and E-business Related Research, Sinteza: 293–8.

Wang, C., Tianhong, L. and Lijun, Z. 2019. Window zooming-based localization algorithm of fruit and vegetable for harvesting robot. IEEE Access. 7: 103639–49.

Wang, J., Tchapmi, L. P., Ravikumar, A. P. and MacGuire, M. 2020. Machine vision for natural gas methane emissions detection using an infrared camera. Applied Energy. 257: 1–10.

Zancul, E., Martins, H. O. and Lopes, F. P. 2020. Machine vision applications in a learning factory machine vision applications in a learning factory. Procedia Manufacturing. 45: 516–21. https://doi.org/10.1016/j.promfg.2020.04.069.

Zhang, J., Ding, G., Zou, Y., Qin, S. and Fu, J. 2017. Review of job shop scheduling research and its new perspectives under Industry 4.0. Journal of Intelligent Manufacturing, 30(3): 1–22.

Zheng, P., Wang, H., Sang, Z., Zhong, Z. Y., Liu, Y., Liu, C., Mubarok, K., Yu, S. and Xu, X. 2018. Smart manufacturing systems for Industry 4.0: Conceptual framework, scenarios, and future perspectives. Frontiers of Mechanical Engineering. 13(2): 137–50.

Zhou, Q., Chen, R., Huang, B. and Liu, C. 2019. An automatic surface defect inspection system for automobiles using machine vision methods. Sensors: 1–18. doi: 10.3390/s19030644.

Zhu, Z., German, S. and Brilakis, I. 2011. Visual retrieval of concrete crack properties for automated post- earthquake structural safety evaluation. Automation in Construction. 20(7): 874–83.

14 Industry 5.0
The Integration of Modern Technologies

Mahadi Hasan Miraz, Mohammad Tariq Hasan, Farhana Rahman Sumi, Shumi Sarkar and Mohammad Amzad Hossain

CONTENTS

DOI: 10.1201/9781003122401-14

14.1 INTRODUCTION

The manufacturing industry is intelligent manufacturing, Industry 4.0 and cyber-physical structures (Collins, 2020; Whittle, 2019; Zhang, Yan, & Wen, 2020). This article is intended to give a sense of specific vital topics that are essential if the future of production is to be successfully carried out (Al Faruqi, 2019). This study includes a hybrid system, advanced building blocks for development, simultaneous technological maturation, production, market situation and accessibility to the environment (Aslam *et al.*, 2020; Bagheri *et al.*, 2015). Improve technical scalability and collaboration between industry and academia, as well. This study included the integration of swarm robotics in modern Industry 5.0.

Swarm robotics is a coordination method and a system consisting of many physical robots (Banjanović-Mehmedović & Mehmedović, 2020; Bousdekis, Apostolou, & Mentzas, 2020). IT is expected to emerge as the desired collective action from the interactions between robots' interactions with the environment. The cooperative behavior of artificial, autonomous and self-organised systems is Swarm Intelligence (SI). The concept used for industry in artificial intelligence 5.0 (Chen, 2017; Clark *et al.*, 2020).

14.1.1 PRACTICAL PROBLEMS IN THE ARENA OF MACHINE VISION AND INDUSTRY 4.0

Machine vision is now making an essential contribution to the automotive industry, mainly as part of QC procedures by offering automatic inspection capability. The automation universe, however, is becoming increasingly complicated. Industry 4.0, the Internet of Things (IoT), cloud computing, artificial intelligence, computer learning and many other innovations provide consumers and operators of vision systems with huge obstacles in choosing the ideal device for their respective applications.

- *Enabling Technology*
  Critically, Industry 4.0 requires a standard communication protocol for all sensor types to allow data transfer and sharing.
- *Aiding Manual Assembly*
  There are also large quantities of manually designed items, and a "human assist" camera should be used to help avoid mistakes in those tasks.
- *Adding Vision to the Production Line*
  It's a well-established method to use vision inspection on a production or packaging line. Systems range from single-point self-contained smart cameras that conduct an inspection role and provide the control system with

a pass/fail result to PC-based systems that can feature multiple cameras and multiple inspection stations.

- *Vision-Guided Robots*

    Industrial robots are already used extensively and are being used even more in combination, particularly for vision-guided robotics, with the advent of collaborative robots and rapid advances in 3D image processing.

- *Machine and Deep Learning*

    Machine vision regarding deep learning uses convolutionary neural networks (CNNs) to perform classification tasks by recognising features learned from a collection of training images.

14.1.2 MANUFACTURING REDEFINITION

Manufacturing contributes to the national economy, including the GDP (Davies, Coole, & Smith, 2017; Demir, Döven, & Sezen, 2019). Policymakers and the general public need to consider production's effect on the country's economy, culture and economic portfolios (Connolly-Barker *et al.*, 2020; Davidson, 2020). However, raising public awareness and winning support from politicians is not always an easy job (Derigent, Cardin, & Trentesaux, 2020; Egger & Masood, 2020). One issue is that most individuals are unfamiliar with the product image (Wang *et al.*, 2020). Redefining the development of future manufacturing is the first significant challenge. Another is to educate the general public about how our economy and society are affected by output. In addition policymakers' unambiguous support is required (Fantini, Pinzone, & Taisch, 2020; Ge *et al.*, 2020). Modern manufacturing processes, emerging technologies and revolutionary business models will dramatise our knowledge base. We also need to create what we consider major challenges (Guo *et al.*, 2020). In addition, additive technology in development will improve manufacturing production (Guo *et al.*, 2020; Guo *et al.*, 2020).

Because of the increased availability of materials, properties, performance and quality, additive production is more widely accepted as a "direct manufacturing" process (Haleem & Javaid, 2019). The output of additives should not be treated as a simple process of manufacturing. This reveals further improvements in productivity and changes in how products are manufactured and delivered (He & Bai, 2020). The GE90 jet engine fuel nozzle is an example of a well-documented and useful contribution to the consolidation of additive development components. The fuel tube incorporates into a single unit all 20 elements of the ancient design but weighs 25% less. Using an additive manufacturing process is more than five times greater (Hodgkins, 2020). The group is much less susceptible to collection sizes through additional production (Horick, 2020).

Manufactured services are becoming the main engine of value production concerning the business model. Output initiates and enables technology to create value in these instances (Jian-zhong, 2014). Many manufacturers have recognised that production and service overlap because the environment has shifted from product delivery to continued customer interaction (Hou *et al.*, 2020). For many producers of creative and cost-effective services, intelligent sensors and communications are rapidly becoming a business model (Krugh & Mears, 2018; Kusiak, 2017). More efficient mixed business models for manufacturing services are available (Hozdić & Butala, 2020). Rolls-Royce, for instance, uses sensors in its jet engines to track performance better and identify problems. However, this business turns the product into service by charging customers for engine use rather than allowing customers to purchase a generator (Jeon et al., 2020; Jeyanthi, 2018). As a further example, Babolat developed sensor rackets for a tennis player to generate information that allows a company to provide coaching services (Lee, Azamfar, & Singh, 2019).

14.1.3 Hybrid Manufacturing

About three decades ago, the invention led to the use of robotics in factories. In ten years, people expected all factories to be filled with robots, and there would be no human workers (Lee et al., 2018; Li et al., 2017). Human operators are working in factories decades later and will continue in the foreseeable future (Li et al., 2020). A network of hybrid manufacturing systems and machinery will introduce a new role. The technology utilises additional and subtractive substances, composites and metals, cyber and physical structures, nano and macro scales, etc. (Li et al., 2020). The production of additives will not completely replace humans, as will the output of subtractions. Instead, they work together to assign tasks equally. Research into individual manufacturing structures is essential (Lins & Oliveira, 2020), also necessary to operate the hybrid system effectively and effectively (Ma et al., 2020).

Shape, machining (subtractive) and additive production are the three main manufacturing groups (Maresova et al., 2018). A graph shows the relationship between unit cost and output volume (or lot size). Shaping is ideal for large lots, with the amortisation of many initial capital expenses in equipment and machinery, to achieve low unit costs (Moore, 2020). The batch volume for processing is smaller in shape than that for additive manufacturing (Nahavandi, 2019; O'donovan et al., 2018).

It is essential to mention that the development point between subtractive and additive production matures as direct production (Özdemir & Hekim, 2018). This makes additive production more cost-effective with new uses (Ozkeser, 2018; Pacaux-Lemoine et al., 2017). Similar technical and economic analysis should be carried out concerning human-robot integration and composite-metal integrity (Paschek, Mocan, & Draghici, 2019).

14.2 BUILDING BLOCKS OF THE FUTURE OF MANUFACTURING ENTERPRISES

These past industrial revolutions took at least 80 years, and Industry 4.0 remains early to be described (Pires, Barbosa, & Leitão, 2018; Preuveneers & Ilie-Zudor, 2017). To the best understanding, it is fair to think that there are active manufacturing companies.

14.2.1 Digital Threads

The security community originally developed the concepts of digital twins and digital strings, which are now the developers' initial community. Digital doublers provide information on the asset's activities, including design requirements, technical models, support and operating specifications (Sachsenmeier, 2016).

14.2.2 SCM Traceability

For modern manufacturing architectures, the recognition of device technology openness is strongly recommended (Miraz, Hasan, & Sharif, 2018; Miraz *et al.*, 2020; Miraz *et al.*, 2019). It goes beyond a manufacturing or distribution center to optimise the entire supply chain's optimisation or customer as the whole base (Saldivar *et al.*, 2015). We are one major step closer to full accountability with the IoT and public threads and visibility in the supply chain.

14.2.3 Hybrid Manufacturing

The focus of future production is on hybrid systems. The hybrid system is also addressed in the previous sections.

14.2.4 Advanced Resources

Raw material, time, stone, bronze and iron instruments are advance production required to manufacture. The focus on material innovation must continue to be on science, technology and economic policy. The production of materials is inseparable (Saldivar *et al.*, 2015).

14.2.5 Innovative Metrology

As a general rule of innovation metrology, the order of the dimension is more significant than the average (Shan *et al.*, 2020). Nano-processing also becomes a routine technique in specific sectors, demand for more advanced metrology.

14.2.6 An Expert Staff for IS

Two elements of this deficiency are present. A shortage of professional manufacturing workers may be the most significant barrier to the intellectual development of the next decade in the world. Second, it is not sufficient for individuals with the skills required to fill production roles (Shi *et al.*, 2020). In other words, inadequate quantities of labor and productivity are used to generate the future. Manufacturing operators will be in short supply of 3.5 million in the United States alone by 2030 (Skobelev & Borovik, 2017).

14.2.7 Ultramodern Business Model

As discussed earlier in this article, we see an increase in manufacturing and operation in many fields. We're coming up with questions from the research.

14.3 RESEARCH QUESTIONS

1. How can the Industry 5.0 model be developed in the manufacturing industry?
2. How to assess the proposed model of requirements for Industry 5.0?

14.4 INDUSTRY REVOLUTION

A comparative summary of industrial revolutions' history shows that each stage lasted from 80 to 100 years. This means that a transition will not occur immediately, but rather that gradual changes and improvements will culminate (Smith, 2020). The Fourth Industrial Revolution is at its beginning and is expected to be at least run for 50 years (Sun *et al.*, 2020). The variables that define every industrialisation process are given below (Tao *et al.*, 2019). Industry 2.0 is characterised by standardisation and easy functional automation, including the standardisation of the basis for mass production and automation. In several inspections and quality controls, unification takes place (Miraz, 2020; Miraz & Habib, 2016; Miraz, Habib, & Molla, 2016). Simple pick-and-place and component feeders based on vibration are just examples of simple hardwired automation (Tao *et al.*, 2019). The objective is to increase hardwired devices' automation (Villalonga *et al.*, 2020). Process performance is a significant concern in the creation of systems. Industry 3.0 is defined by improving automation, digitisation and networking (Shi *et al.*, 2020; Skobelev & Borovik, 2017). With speed, productivity and versatility in processing, this step provides much more advanced automation.

Advanced robotics and programming are the quintessential class of scalable automation (Miraz, Hasan, & Sharif, 2019; Miraz *et al.*, 2020; Miraz *et al.*, 2019). To accommodate product variability and the size fluctuation of the lot, the supplier should provide appropriate response and precision with versatile automation (Villalba-Díez *et al.*, 2020). Other characteristic features of Industry 3.0 and the collection of data are the instruments used in machinery (number control devices, 3D printers and robots) with sensors to collect data for process monitoring, control and management. Industry 4.0, the last element which distinguishes Industry 4.0 and Industry 3.0, is supported by different technology networks (Li *et al.*, 2020; Lins & Oliveira, 2020; Nahavandi, 2019; Saldivar *et al.*, 2015).

Computer-based factories communicate in real-time through their network of subnetworks with each other (Zhou, Liu, & Zhou, 2015). Sensors, data sharing and networking provide unparalleled strength in industrial and manufacturing enterprises. Nevertheless, they also face these companies' security problems (Zhou *et al.*, 2019). The investigator, therefore, conceptualised the concept of 5.0.0. And what is the difference between Industry 5.0 and Industry 4.0? How does Industry 5.0 move above and beyond Industry 4.0 to achieve this?

Here are a couple of thoughts.

14.4.1 EXTERNAL OPTIMISATION

The distribution is crucial for industrial and manufacturing companies to optimise output/input ratios. Optimisation is essential for Industry 4.0, but not enough. In other words, optimisation alone is not enough to distinguish between companies.

14.4.2 SITUATIONAL AWARENESS

All production systems are composed of machine tools, robots and 3D printers. It must monitor the environment and make individual decisions (Zhou *et al.*, 2018). Like in Industry 4.0, most decisions are made based on preliminary information and historical evidence. An Industry 5.0 organisation, however, should be able to identify the unidentified scenario. It also needs to break down the issue into bits, use technology to resolve the problems and search for information and facts. Once the "latest" or "unique" problems have been resolved, the system will become "more intelligent" than before because lessons have been learned from the machine (Al Faruqi, 2019; Banjanović-Mehmedović & Mehmedović, 2020; Bousdekis *et al.*, 2020; Demir *et al.*, 2019; Derigent *et al.*, 2020; Haleem & Javaid, 2019). Cognitive awareness and listening abilities are perhaps the most distinctive features of Industry 5.0 (Zhong *et al.*, 2017; Zhou *et al.*, 2020).

14.4.3 DESIGNED AND FORMLESS DATA

The Industry 5.0 standard is the extensive use of structured and nonstructured data. Solutions are no longer limited to the structured data domain (Zheng, Xu, & Chen, 2020). Smart solutions offer unstructured data sets such as images, natural languages, even social media and global market updates.

14.4.4 PERFORMANCE METRICS

Furthermore, new variables such as resilience, adaptability and the ability to learn from failure or human behavior have been identified. In the Industry 5.0 performance indicators, such as efficiency, quality, replicability, costs and risk, performance will be more critically integrated (Villalba-Díez *et al.*, 2020; Villalonga *et al.*, 2020).

Figure 14.1 shows the industry transformation from 1 to 4. Table 14.1 describes Industry 5.0 literature described technologies.

Industry 1.0	Industry 2.0	Industry 3.0	Industry 4.0
Mechanization	Mass production	Automation	Cyber-physical systems
Steam power	Assembly line	Computers	Internet of things
Weaving loom	Electrical energy	Electronics	Networks

FIGURE 14.1   Industries 1.0 to 4.0.

TABLE 14.1
Industry 5.0 Literature Described Technologies

No.	Technology	No. Technology Definition
1	IIoT (Industrial Internet of Things)	Real-time associated with human beings, machines, objects and ICT systems able to handle complex systems smartly, horizontally and vertically
2	CPS (cyber-physical systems)	Creating global business networks to merge the physical and digital worlds
3	Big data collection and analysis	Extract data from large amounts to make educated decisions
4	Cloud services for products	To use cloud computing in goods to expand their functionality and related services, collect information from enormous volumes of data
5	Additive manufacturing	Compare with conventional machines such as lathes or frying machines for removal of chips, e.g. printing in 3D
6	Simulations/analysis of virtual models	Virtual model analysis using finite element analysis, CFD, in which simulations simulate the properties of applied simulations dynamics
7	Integrated engineering systems	In product development and production, IT support systems are incorporated for knowledge exchange
8	Augmented reality	A technique in computer graphics that superposes simulated signals to a real image of the outside world
9	Flexible manufacturing lines	Digital automation with sensors, for example, by RFID or the development of processes. Reconfigurable manufacturing systems (RMS)
10	Cybersecurity	Protection against cyberattacks of Internet-linked devices such as data information, hardware and software
11	Swarm robotics	The field of swarm robotics is a multi-robotic field in which many robots are distributed and decentralised. It is based on the application of local rules and simple robots compared to the task's difficulty and inspired by insects

14.5   CONCEPTUAL FRAMEWORK

A few more variables were incorporated into Industry 5.0.0 by the investigator. The scientists introduce the new robotic facility, smart contract, blockchain and swarm as a new variable that improves Industry 5.0. Figure 14.2 shows the conceptual framework of this research.

14.5.1   METHODOLOGY

The theoretical research is carried out by analysing the literature to define the problems and gaps associated with the study's domain. Consequently, by reading printed and online references, the main concepts were obtained through the literature. In addition, in the industry's expert opinion, the current problems of Industry 4.0 and its limitations are recognised. Once the information has been collected, an analysis is carried out on the existing manufacturing sector, which will lead to the implementation factors of Industry 5.0 in the manufacturing industry.

14.5.2   THE BUILDING OF INDUSTRY 5.0 FOR THE MANUFACTURING INDUSTRY

In this phase, using the identified parameters, the development of an Industry 5.0 will be carried out. For the manufacturing industry, particular steps have been developed. There are three levels involved: development of industrial 5.0, quantification of 5.0 components and robotic system Swarm.

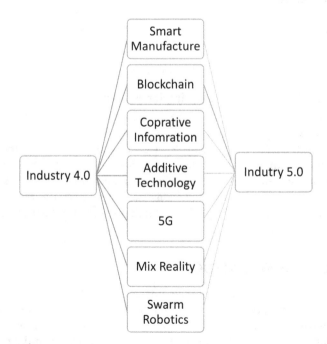

FIGURE 14.2   Conceptual framework of this research.

14.5.3 Creation of Requirement Model

The integrated approach will be developed as a result of the Phase 2 phase. In this research, Phase 3 is the critical phase. This stage is made up of several steps. These include Industry 5.0 factor analysis, Industry 5.0 impact and the robotic system Swarm.

A modeling tool such as BUZZ programming language to construct a swarm base system during this phase. Buzz, a modern robot swarm programming language, is presented by the investigator. Buzz supports a compositional approach that provides essential elements for both the robot and the swarm as a whole to describe swarm behavior. Robot-specific directions are supplied by primitive single robot guidance and handling of neighborhood information. Swarm-based primitives enable dynamic robot management and worldwide sharing of knowledge within the swarm. Self-organisation arises from the completely decentralised processes that form the foundation of the Buzz platform for runtime. It is also possible to extend the vocabulary to include new primitives of heterogeneous robot swarms that can be used on other systems, including the robot's operating system. It has an interface with working time. The skills of Buzz are demonstrated by code presentation and realistically simulated experiments on typical swarm algorithms.

14.5.4 Construction of the Requirement Model Using UML

In various techniques, case studies and expert reviews, the proposed integrated model was assessed and analysed.

14.5.4.1 Case Study

The anticipated integration is applied and validated at the selected organisation—the case study conducted for testing and assessment.

14.5.4.2 Professional Reviews to Implement 5.0

In line with the study findings, industry expert reviews were carried out to validate the model proposed to build qualitative analyses to implement Industry 5.0. Based on their experience and experience in Industry 4.0 in the manufacturing industry, experts are selected. An expert's feedback serves as an input for improving and improving the integrated model proposed by the industry. Compared with the findings of the industrial experts are the results from the developed model and case study. The research process of Industry 5.0, therefore, enforces achievement.

14.6 BODY OF KNOWLEDGE

The research mainly contributes to new information or knowledge to enhance the improvement of products through Industry 5.0. It can improve industrial production and an easy procedure for supply chain management. Moreover, manufacture and consumer have excellent dealing and cooperative behavior in the national and international industries.

14.6.1 PROSPECTIVE CONSUMERS

In the world, modern manufacturing has been promoted as a particular manufacturing industry. It has a reputation for serving customers from all around the world. In addition, customers can find easy access to industrial production.

14.6.2 TO THE INVESTORS

Industry 5.0 offers the worldwide manufacturing industry the ultimate solution. Industry 5.0 emphasises the manufacturing industry's local and rural businesses. The business strategies will be altered, and the live hood will be improved through Industry 5.0.

14.6.3 FINDINGS/KNOWLEDGE

This model introduces Industry 5.0 from an industrial perspective, which is to be implemented in the manufacturing industry's production. The model offers a variety of scenarios by allowing producers to alter the model data. The industry can develop data factors to select a more efficient system for future manufacture. It will also create a feeling for future production. The most efficient use of a company's resources can be achieved by changing the model's variables. This technique was developed by awareness of the importance of managing the supply chain operation of the industry. Therefore, this research has contributed to precisely automated and virtual production in the manufacturing industry.

14.7 CONCLUSION

The researcher proposed a model for Industry 5.0.0 in this study. Information and aspects relating to system architecture, design, diagrams of entity-relation, interactions and algorithms for implementation have been presented. It is our solution to monitor and monitor the industry's supply chain. However, the aspects and information provided are sufficiently standardised and can give precise and decentralised traceability for any product produced in the industry's supply chain. The main challenges of scalability, governance, identity registration, protection, standards, and regulations are also faced by Industry 5.0.

REFERENCES

Al Faruqi, U. (2019). Future service in Industry 5.0. *Jurnal Sistem Cerdas*, 2(1), 67–79.
Aslam, F., Aimin, W., Li, M., & Ur Rehman, K. (2020). Innovation in the era of IoT and Industry 5.0: Absolute innovation management (AIM) framework. *Information*, 11(2), 124.
Bagheri, B., Yang, S., Kao, H.-A., & Lee, J. (2015). Cyber-physical systems architecture for self-aware machines in industry 4.0 environment. *IFAC-PapersOnLine*, 48(3), 1622–1627.
Banjanović-Mehmedović, L., & Mehmedović, F. (2020). Intelligent manufacturing systems driven by artificial intelligence in industry 4.0. In *Handbook of Research on Integrating Industry 4.0 in Business and Manufacturing* (pp. 31–52): USA: IGI Global.

Bousdekis, A., Apostolou, D., & Mentzas, G. (2020). A human cyber physical system framework for operator 4.0—Artificial intelligence symbiosis. *Manufacturing Letters*, *25*, 10–15.

Chen, Y. (2017). Integrated and intelligent manufacturing: Perspectives and enablers. *Engineering*, *3*(5), 588–595.

Clark, A., Zhuravleva, N. A., Siekelova, A., & Michalikova, K. F. (2020). Industrial artificial intelligence, business process optimization, and big data-driven decision-making processes in cyber-physical system-based smart factories. *Journal of Self-Governance and Management Economics*, *8*(2), 28–34.

Collins, K. (2020). Cyber-physical production networks, real-time big data analytics, and cognitive automation in sustainable smart manufacturing. *Journal of Self-Governance and Management Economics*, *8*(2), 21–27.

Connolly-Barker, M., Gregova, E., Dengov, V. V., & Podhorska, I. (2020). Internet of Things sensing networks, deep learning-enabled smart process planning, and big data-driven innovation in cyber-physical system-based manufacturing. *Economics, Management and Financial Markets*, *15*(2), 23–29.

Davidson, R. (2020). Cyber-physical production networks, artificial intelligence-based decision-making algorithms, and big data-driven innovation in industry 4.0-based manufacturing systems. *Economics, Management, and Financial Markets*, *15*(3), 16–22.

Davies, R., Coole, T., & Smith, A. (2017). Review of socio-technical considerations to ensure successful implementation of Industry 4.0. *Procedia Manufacturing*, *11*, 1288–1295.

Demir, K. A., Döven, G., & Sezen, B. (2019). Industry 5.0 and human-robot co-working. *Procedia Computer Science*, *158*, 688–695.

Derigent, W., Cardin, O., & Trentesaux, D. (2020). Industry 4.0: Contributions of holonic manufacturing control architectures and future challenges. *Journal of Intelligent Manufacturing*, *32*, 1–22.

Egger, J., & Masood, T. (2020). Augmented reality in support of intelligent manufacturing—A systematic literature review. *Computers & Industrial Engineering*, *140*, 106195.

Fantini, P., Pinzone, M., & Taisch, M. (2020). Placing the operator at the centre of Industry 4.0 design: Modelling and assessing human activities within cyber-physical systems. *Computers & Industrial Engineering*, *139*, 105058.

Ge, J., Wang, F., Sun, H., Fu, L., & Sun, M. (2020). Research on the maturity of big data management capability of intelligent manufacturing enterprise. *Systems Research and Behavioral Science*, *37*(4), 646–662.

Guo, D., Zhong, R. Y., Lin, P., Lyu, Z., Rong, Y., & Huang, G. Q. (2020). Digital twin-enabled Graduation Intelligent Manufacturing System for fixed-position assembly islands. *Robotics and Computer-Integrated Manufacturing*, *63*, 101917.

Guo, D., Zhong, R. Y., Ling, S., Rong, Y., & Huang, G. Q. (2020). A roadmap for Assembly 4.0: Self-configuration of fixed-position assembly islands under Graduation Intelligent Manufacturing System. *International Journal of Production Research*, *5*,1–16.

Guo, Y., Wang, N., Xu, Z.-Y., & Wu, K. (2020). The Internet of Things-based decision support system for information processing in intelligent manufacturing using data mining technology. *Mechanical Systems and Signal Processing*, *142*, 106630.

Haleem, A., & Javaid, M. (2019). Industry 5.0 and its applications in orthopaedics. *Journal of Clinical Orthopaedics and Trauma*, *10*(4), 807–808.

He, B., & Bai, K.-J. (2020). Digital twin-based sustainable intelligent manufacturing: A review. *Advances in Manufacturing*, *9*,1–21.

Hodgkins, S. (2020). Cyber-physical production networks: Artificial intelligence data-driven internet of things systems, smart manufacturing technologies, and real-time process monitoring. *Journal of Self-Governance and Management Economics*, *8*(1), 114–120.

Horick, C. (2020). Industry 4.0 production networks: Cyber-physical system-based smart factories, real-time big data analytics, and sustainable product lifecycle management. *Journal of Self-Governance and Management Economics*, *8*(1), 107–113.

Hou, T., Cheng, B., Wang, R., Xue, W., & Chaudhry, P. E. (2020). Developing Industry 4.0 with systems perspectives. *Systems Research and Behavioral Science, 37*(4), 741–748.

Hozdić, E., & Butala, P. (2020). Concept of socio-cyber-physical work systems for Industry 4.0. *Tehnički vjesnik, 27*(2), 399–410.

Jeon, B., Yoon, J.-S., Um, J., & Suh, S.-H. (2020). The architecture development of Industry 4.0 compliant smart machine tool system (SMTS). *Journal of Intelligent Manufacturing, 31*(8), 1–23.

Jeyanthi, P. M. (2018). Industry 4.0: The combination of the Internet of Things (IoT) and the Internet of People (IoP). *Journal of Contemporary Research in Management, 13*(4), 1–25.

Jian-zhong, F. (2014). Development status and trend of intelligent manufacturing equipment. *Journal of Mechanical & Electrical Engineering, 31*(8), 959–962.

Krugh, M., & Mears, L. (2018). A complementary cyber-human systems framework for industry 4.0 cyber-physical systems. *Manufacturing Letters, 15*, 89–92.

Kusiak, A. (2017). Smart manufacturing must embrace big data. *Nature, 544*(7648), 23–25.

Lee, J., Azamfar, M., & Singh, J. (2019). A blockchain enabled cyber-physical system architecture for Industry 4.0 manufacturing systems. *Manufacturing Letters, 20*, 34–39.

Lee, J., Davari, H., Singh, J., & Pandhare, V. (2018). Industrial artificial intelligence for industry 4.0-based manufacturing systems. *Manufacturing Letters, 18*, 20–23.

Li, B.-H., Hou, B.-C., Yu, W.-T., Lu, X.-B., & Yang, C.-W. (2017). Applications of artificial intelligence in intelligent manufacturing: A review. *Frontiers of Information Technology & Electronic Engineering, 18*(1), 86–96.

Li, X., Wang, B., Liu, C., Freiheit, T., & Epureanu, B. I. (2020). Intelligent manufacturing systems in COVID-19 pandemic and beyond: Framework and impact assessment. *Chinese Journal of Mechanical Engineering, 33*(1), 1–5.

Lins, T., & Oliveira, R. A. R. (2020). Cyber-physical production systems retrofitting in context of industry 4.0. *Computers & Industrial Engineering, 139*, 106193.

Ma, S., Zhang, Y., Liu, Y., Yang, H., Lv, J., & Ren, S. (2020). Data-driven sustainable intelligent manufacturing based on demand response for energy-intensive industries. *Journal of Cleaner Production, 274*, 123155.

Maresova, P., Soukal, I., Svobodova, L., Hedvicakova, M., Javanmardi, E., Selamat, A., & Krejcar, O. (2018). Consequences of industry 4.0 in business and economics. *Economies, 6*(3), 46.

Miraz, M. (2020). Blockchain in automotive supply chain. *International Supply Chain Technology Journal, 6*(6), 1–12. doi:10.20545/isctj.v06.i06.02

Miraz, M. H., & Habib, M. M. (2016). Effect of information technology in the automotive supply chain. *Open Journal of Technology & Engineering Disciplines (OJTED), 2*(1), 28–32.

Miraz, M. H., Habib, M. M., & Molla, M. S. (2016). An overview of information technology tools implementation in supply chain management. *IETI Transactions on Computers, 2*(2), 110–117.

Miraz, M. H., Hasan, M. G., & Sharif, K. I. (2018). The relationship between personal and organizational in supply chain integration: Case study in Malaysia. *Journal of Business Management and Economic Research, 2*(7), 43–48. doi:10.29226/tr1001.2018.48

Miraz, M. H., Hasan, M. G., & Sharif, K. I. (2019). The numerous tactical plans affect customer and postal service relationship: The mediating role of blockchain, an empirical study in Bangladesh. *Journal of Advanced Research in Dynamical & Control Systems, 11*(5), 985–990.

Miraz, M. H., Hasan, M. T., Sumi, F. R., Sarkar, S., & Majumder, M. I. (2020). The innovation of blockchain transparency & traceability in logistic food chain. *International Journal of Mechanical and Production Engineering Research and Development (IJMPERD)*, *10*(3), 9155–9170.

Miraz, M. H., Hye, A. K. M., Wahab, M. K., Alkurtehe, K. A. M., Majumder, M. I., Habib, M. M., & Alsabahi, M. A. (2020). Blockchain securities to construct inclusive, digital economy globally. *International Supply Chain Technology Journal*, *6*(1), 1–11. doi:10.20545/isctj.v06.i01.03

Miraz, M. H., Kabir, A., Habib, M. M., & Alam, M. M. (2019). Blockchain technology in transport industries in Malaysia. *Paper presented at the 2nd international conference on business and management.*

Moore, K. (2020). Smart connected sensors, cyber-physical networks, and big data analytics systems in internet of things-based real-time production logistics. *Economics, Management, and Financial Markets*, *15*(2), 16–22.

Nahavandi, S. (2019). Industry 5.0—A human-centric solution. *Sustainability*, *11*(16), 4371.

O'donovan, P., Gallagher, C., Bruton, K., & O'Sullivan, D. T. (2018). A fog computing industrial cyber-physical system for embedded low-latency machine learning Industry 4.0 applications. *Manufacturing Letters*, *15*, 139–142.

Özdemir, V., & Hekim, N. (2018). Birth of industry 5.0: Making sense of big data with artificial intelligence, "the internet of things" and next-generation technology policy. *Omics: A Journal of Integrative Biology*, *22*(1), 65–76.

Ozkeser, B. (2018). Lean innovation approach in Industry 5.0. *The Eurasia Proceedings of Science, Technology, Engineering & Mathematics*, *2*, 422–428.

Pacaux-Lemoine, M.-P., Trentesaux, D., Rey, G. Z., & Millot, P. (2017). Designing intelligent manufacturing systems through human-machine cooperation principles: A human-centered approach. *Computers & Industrial Engineering*, *111*, 581–595.

Paschek, D., Mocan, A., & Draghici, A. (2019). Industry 5.0—The expected impact of next industrial revolution. *Paper presented at the thriving on future education, industry, business, and society, proceedings of the MakeLearn and TIIM international conference, Piran, Slovenia.*

Pires, F., Barbosa, J., & Leitão, P. (2018). Quo vadis Industry 4.0: An overview based on scientific publications analytics. *Paper presented at the 2018 IEEE 27th international symposium on industrial electronics (ISIE).*

Preuveneers, D., & Ilie-Zudor, E. (2017). The intelligent industry of the future: A survey on emerging trends, research challenges and opportunities in Industry 4.0. *Journal of Ambient Intelligence and Smart Environments*, *9*(3), 287–298.

Sachsenmeier, P. (2016). Industry 5.0—The relevance and implications of bionics and synthetic biology. *Engineering*, *2*(2), 225–229.

Saldivar, A. A. F., Li, Y., Chen, W.-N., Zhan, Z.-H., Zhang, J., & Chen, L. Y. (2015). Industry 4.0 with cyber-physical integration: A design and manufacture perspective. *Paper presented at the 2015 21st international conference on automation and computing (ICAC).*

Shan, S., Wen, X., Wei, Y., Wang, Z., & Chen, Y. (2020). Intelligent manufacturing in industry 4.0: A case study of Sany Heavy Industry. *Systems Research and Behavioral Science*, *37*(4), 679–690.

Shi, Z., Xie, Y., Xue, W., Chen, Y., Fu, L., & Xu, X. (2020). Smart factory in Industry 4.0. *Systems Research and Behavioral Science*, *37*(4), 607–617.

Skobelev, P., & Borovik, S. Y. (2017). On the way from Industry 4.0 to Industry 5.0: From digital manufacturing to digital society. *Industry 4.0*, *2*(6), 307–311.

Smith, A. (2020). Cognitive decision-making algorithms, real-time sensor networks, and internet of things smart devices in cyber-physical manufacturing systems. *Economics, Management, and Financial Markets*, *15*(3), 30–36.

Sun, Y., Li, L., Shi, H., & Chong, D. (2020). The transformation and upgrade of China's manufacturing industry in Industry 4.0 era. *Systems Research and Behavioral Science, 37*(4), 734–740.

Tao, F., Qi, Q., Wang, L., & Nee, A. (2019). Digital twins and cyber–physical systems toward smart manufacturing and industry 4.0: Correlation and comparison. *Engineering, 5*(4), 653–661.

Villalba-Díez, J., Molina, M., Ordieres-Meré, J., Sun, S., Schmidt, D., & Wellbrock, W. (2020). Geometric deep lean learning: Deep learning in industry 4.0 cyber–physical complex networks. *Sensors, 20*(3), 763.

Villalonga, A., Beruvides, G., Castaño, F., & Haber, R. (2020). Cloud-based industrial cyber-physical system for data-driven reasoning. A review and use case on an industry 4.0 pilot line. *Statistics, 34*, 35.

Wang, B., Hu, S. J., Sun, L., & Freiheit, T. (2020). Intelligent welding system technologies: State-of-the-art review and perspectives. *Journal of Manufacturing Systems, 56*, 373–391.

Whittle, T. (2019). Interaction networks in the production and operations environment: Internet of things, cyber-physical systems, and smart factories. *Journal of Self-Governance and Management Economics, 7*(2), 13–18.

Zhang, H., Yan, Q., & Wen, Z. (2020). Information modeling for cyber-physical production system based on digital twin and AutomationML. *The International Journal of Advanced Manufacturing Technology, 5*(4)1–19.

Zheng, P., Xu, X., & Chen, C.-H. (2020). A data-driven cyber-physical approach for personalized smart, connected product co-development in a cloud-based environment. *Journal of Intelligent Manufacturing, 31*(1), 3–18.

Zhong, R. Y., Xu, X., Klotz, E., & Newman, S. T. (2017). Intelligent manufacturing in the context of industry 4.0: A review. *Engineering, 3*(5), 616–630.

Zhou, G., Zhang, C., Li, Z., Ding, K., & Wang, C. (2020). Knowledge-driven digital twin manufacturing cell towards intelligent manufacturing. *International Journal of Production Research, 58*(4), 1034–1051.

Zhou, J., Li, P., Zhou, Y., Wang, B., Zang, J., & Meng, L. (2018). Toward new-generation intelligent manufacturing. *Engineering, 4*(1), 11–20.

Zhou, J., Zhou, Y., Wang, B., & Zang, J. (2019). Human–cyber–physical systems (HCPSs) in the context of new-generation intelligent manufacturing. *Engineering, 5*(4), 624–636.

Zhou, K., Liu, T., & Zhou, L. (2015). Industry 4.0: Towards future industrial opportunities and challenges. *Paper presented at the 2015 12th International conference on fuzzy systems and knowledge discovery (FSKD).*

Index

CPSIA information can be obtained
at www.ICGtesting.com
Printed in the USA
BVHW091744190422
634676BV00002B/23